Sex Wars

Sex Wars

GENES, BACTERIA, AND BIASED SEX RATIOS

MICHAEL E. N. MAJERUS

PRINCETON UNIVERSITY PRESS

PRINCETON AND OXFORD

Published by Princeton University Press, 41 William Street, Princeton, New Jersey 08540
In the United Kingdom: Princeton University Press, 3 Market Place, Woodstock, Oxfordshire OX20 1SY

Library of Congress Cataloging-in-Publication Data

Majerus, M. E. N.
Sex wars : genes, bacteria, and biased sex ratios / Michael E.N. Majerus.
p. cm.
Includes bibliographical references and index.
ISBN 0-691-00981-3 (cloth : alk. paper)
1. Sex ratio. 2. Host-bacteria relationships. 3. Invertebrates—Reproduction.
4. Invertebrates—Genetics. I. Title.

QH481 .M25 2003
571.8′45—dc21 2002024340

British Library Cataloging-in-Publication Data is available

This book has been composed in Times Roman

Printed on acid-free paper. ∞

www.pupress.princeton.edu

Printed in the United States of America

1 2 3 4 5 6 7 8 9 10

CONTENTS

ILLUSTRATIONS

FIGURES

PLATES *(following page 154)*

Boxes

TABLES

PREFACE

This is a book about sex. Not doing it, but the root of it. Perhaps more correctly, it is a book about the sexes—the differences between males and females, the differences in their interests, and the conflicts that arise between them as a result of those differences. More specifically, it is a book about sex ratios: the proportion of males and females in populations of sexually reproducing species. It is also a book about bugs—not true bugs of the insect order Hemiptera, although they do get a mention, but bugs in the widest sense, from microscopic bugs, like those, such as bacteria, that cause disease, through the whole gamut of so-called primitive animals, the invertebrates. With a few notable exceptions, such as the alleged habit of female praying mantids and black widow spiders eating their mates after, or occasionally during, copulation, few people, except specialist biologists, know much about the sexual conflicts that exist all around us among the insects and other creepy-crawlies that live with us on Earth.

The central subject of this book concerns bacteria and other microorganisms that are symbionts of multicellular hosts and manipulate the reproduction of their hosts to their own ends. Frequently, this manipulation affects the sex ratio of their host. The sex ratio of organisms is rarely the result of chance. Rather, it is a product of a complex set of interacting phenomena, including the ecologies and behaviors of the species concerned, the way sex is determined during development, the underlying genetics of sex, and, frequently, interactions between the animals and microorganisms that live inside them.

The book has two parts. Part I, comprising the first four chapters, is concerned with theories and experimental evidence on sex, sex ratios, sex determination mechanisms, and sexual conflicts. Part II focuses on inherited elements that flout the normal Mendelian rules of inheritance and, in particular, inherited microorganisms that impact on the sex ratios and evolution of their hosts.

The root of sexual conflict is the essential difference between males and females: the size of the sex cells they produce. This difference leads to alternative reproductive strategies in males and females (chapter 1).

In many organisms, approximately equal numbers of males and females are produced. The explanation of why this is so is given in chapter 2. Yet there is a sizable minority of organisms in which more females are produced than males, and a smaller number of cases in which males are produced in excess. Cases of biases in the sex ratio have been known since before Darwin's time. Yet it is only relatively recently, with the development of molecular genetic techniques, that biologists have started to unravel the intricacies of the causes of sex ratio distortion and the reasons why a biased sex ratio may be beneficial, or in some instances severely harmful, to an organism.

Part of the unraveling of sex ratio distortion depends on an understanding of the ways in which sex is determined in different species. Sex determination mechanisms, whether they be genetic, environmental, chemical, mechanical, or any combination of these are the subject of chapter 3.

In most organisms, sex is determined wholly or in part by a gene, or genes, on the chromosomes within the nucleus of cells. Chromosomes are generally inherited equally from the two parents, although the sex chromosomes are an exception in this regard. However, other genetic material is not inherited equally. In animals, most nonnuclear genetic material is inherited only or mainly from the mother. This genetic material should favor the production and survival of females through which it can be inherited, rather than males through which it cannot be. There is, therefore, a conflict between the chromosomes that are inherited from both parents equally and those that are inherited in an unbalanced way. This conflict is explained in chapter 4. Here microorganisms, which live in the cytoplasm of host cells and are inherited maternally, are introduced through the phenomenon of cytoplasmic incompatibility.

The lack of balance in the inheritance of such material has led to the evolution of a number of mechanisms that bias the sex ratio toward females. These include the feminization of males (chapter 5), the killing of males to the benefit of females (chapter 6) and the eradication of the need for males, so that females gain the ability to reproduce without mating (chapter 8).

These mechanisms, in particular the feminization of males and the killing of males, may lead to female-biased population sex ratios. Chapter 7 discusses the evolutionary consequences of sex ratio distortion on the genetics, ecology, and reproductive behavior of the organisms concerned.

The final chapter draws together the threads of material from the proceeding chapters to consider the evolution of sex, why sex is so common, and the conflicts inherent in sex ratio distortion. The implications of the existence of sex ratio distorters are explored and the potential uses of these systems by humans are discussed.

I have tried to use simple terminology where possible, but have had to use some scientific terms for precision and clarity. I have explained these terms on first usage, unless the explanation is obvious from their context. In addition, I have provided a glossary of these terms, as they pertain to their usage in the text. I have referenced the introductory chapters in the book rather sparingly. The later chapters, dealing with the effects of inherited symbionts, are rather more heavily referenced.

In writing this book, my intentions have been twofold. The first is simply to bring some of the amazing examples of the sex wars that go on in the invertebrate world to a wider audience. The cut and thrust of evolutionary arms races that have shaped life on Earth fascinate me, and those that affect sex have always held a special fascination. I hope that readers find some of the stories as spellbinding as I do.

My second intention is to catalyze further work on the sexual conflicts among moths, mites, midges, and a myriad of other small creatures. Research on many of the phenomena discussed in this book is still in its infancy and is conducted by a rather small but expanding number of highly specialized researchers, each with his or her pet organism or two (mine are butterflies, moths, and ladybirds). Sex ratio influences on many groups of invertebrates have never been studied. If in reading this book, readers who have intimate knowledge of a particular group are sparked into some realization that "their organism" may illustrate or confound one or more of the ideas contained herein and so widen the field of investigations into sexual conflicts and sex ratio distortions, I will be delighted.

Cambridge
May 7, 2001

ACKNOWLEDGMENTS

I have a variety of people to thank for their help while I was planning and writing this book. First and foremost are the members of my research group, past and present, whether working on sex ratio distorters or other projects, who have by laboratory work and discussion contributed so much to my views on invertebrate sex-ratios and other sex related phenomena. They are Dominique Bertrand, Dr. Clair Brunton, Marta Chyb, Penny Haddrill, Dr. Greg Hurst, Dr. Francis Jiggins, Tamsin Majerus, Dr. Mark Ransford, Dr. Hinrich Graf von de Schulenburg, Dr. David Shuker, Dr. John Sloggett, Matt Tinsley, and Dr. Mary Weberley. Others with whom I have collaborated, Professor Laurence Hurst, Professor Yoshiaki Obara, Professor Jack Werren, and Professor Ilia Zakharov, have also had an influence. I am also grateful to Dr. Melanie Hatcher, Professor Scott O'Neill, and Dr. Steve Siskins for interesting discussions. The final year undergraduate classes over the last decade at the Department of Genetics in Cambridge have also helped tremendously. Their constructively critical, open-minded attitude to material presented to them has helped me to judge the ease with which various concepts may be understood and which phenomena are of greatest interest to undergraduates.

Much of the illustrative material is my own, but some has come from other workers in the field. I therefore thank for their contributions Ted Bianco, Robert Comte, Alison Dunn, Brij Gupta, Greg Hurst, Francis Jiggins, Tamsin Majerus, Gilbert Martin, Mark Ransford, Hinrich Schulenburg, Richard Stouthamer, Mark Taylor, and Martin Webb.

Throughout the preparation of this book, Sam Elworthy of Princeton University Press has been helpful, encouraging, and patient and has made interaction between author and publisher smooth and surprisingly not time-consuming. I am very grateful to him.

PART I

The Sexes, Sex Determination, and Sex Ratios

CHAPTER 1

The Logic of Sex

> It is interesting to contemplate an entangled bank, clothed with many plants of many kinds, with birds singing on the bushes, with various insects flitting about, and with worms crawling through the damp earth, and to reflect that these elaborately constructed forms, so different from each other, and dependent upon each other in so complex a manner, have all been produced by laws acting around us."
>
> —Charles Darwin, *The Origin of Species*

Summary

In the history of life on Earth, the first organisms reproduced asexually. Sexual reproduction then evolved and came to predominate. In some lineages, some species or some individuals of some species, have reverted to asexual reproduction. The history of changes from asexual to sexual reproduction, and in some cases back again, are outlined.

The predominant form of reproduction—sexual reproduction—involves, in most cases, the existence of two different forms of a species, the male and female sexes. The definer of sex is the size of the reproductive cell produced by an individual. In a species, by definition, males produce smaller reproductive cells than females.

Differences in the size and number of reproductive cells produced by males and females have given the sexes different interests. The differences in the roles of males and females have led to conflicts in relation to reproduction between the sexes. These sexual conflicts have led to various types of sexual selection. Members of one sex, usually males, compete for access to the other sex for fertilization opportunities. Members of the other sex, or in some cases both sexes, exercise careful choice as to which potential mates should be allowed fertilization opportunities. Sexual selection and the ways in which natural selection acts differently on the two sexes have led to a fascinating array of different reproductive strategies. Some of these operate prior to copulation; others after mating has occurred.

Introduction: Fantasy Sex Ratios!

Consider, for a moment, Rincewind, the failed "Wizzard" of Terry Pratchet's Disc World—a world that has kept Pratchet as the top-selling fiction writer in

Britain for over a decade. After being attacked by dragons, tree nymphs, and sorcerers; falling off the world; and accidentally saving it on several occasions, despite his own best efforts just to run away, Rincewind finds his own Nirvana on a desert island. There, he encounters female, "abundantly female," representatives of a tribe who had lost all of their menfolk to a deadly plague, and were seeking a male to enable them to continue their line. I wonder what proportion of men have had a fantasy of this type, and what the female equivalent might be. As a student of evolutionary genetics, I long ago accepted the thesis that the biological roles of males and females are fundamentally different and, consequently, that the Darwinian selection pressures acting on the sexes will differ both in nature and intensity. That being so and being male, I freely admit to having little understanding of women, and so shy away from attempting to offer a female equivalent of Rincewind's male Nirvana.

A strongly female-biased sex ratio has been a feature of many science fiction stories, some of the best known perhaps being episodes of television series such as *Dr. Who*, *Star Trek*, and *Red Dwarf*. In perhaps the most extreme fictional stories, such as John Wyndham's (1965) *Consider Her Ways*, the human population has become solely female, males having been wiped out by a male specific pathogenic virus. As we shall see, this story, originally written in 1956, contains some interesting scientific insights. Not only does it involve a microbial male-killer, but the all-female society that evolves has a caste system, with strong similarities to social insects such as bees, ants, and termites, which are numerically dominated by females, only some of which reproduce. Furthermore, this reproduction is asexual, and involves the production of four offspring at a time, all offspring being female. Again, in this there are hints of some Hymenoptera, which reproduce parthenogenetically, that is, without sex.

Images of female-biased sex ratios are not only found in fantasy or fiction. In some human societies, both past and present, powerful males gather harems of females. Harems are a feature of many other mammals, with males fighting to gain and retain groups of females to mate with. Dominant males who are successful in collecting a harem are under regular attack by subordinate or younger males who try to usurp their position. In these species, the dominant males gain mating opportunities at the expense of other males who are denied them. The sex ratio at birth in such species is not biased; only the sex ratio among reproducing adults is female biased. Yet in some organisms, particularly among the invertebrates, female-biased populations are the norm, and the biases are not confined to reproducing adults.

Reproduction — Asexual, Sexual, and Secondary Asexual

The Ultimate Aim of Existence: The Passage of Genes

It is often assumed that the ultimate aim of an individual is to reproduce. However, there are exceptions. Damaraland mole rats live in large underground colonies, excavating vast networks of tunnels. A colony may consist of up to 40 individuals, all related. All work together to collect tubers and bulbs in the wet season to stock a larder that will sustain the colony through the dry season. Yet, only one pair in the colony breeds. Personal reproduction is put aside in favor of helping siblings. Having the same parents as the other mole rats in the colony, all the individuals share as many genes with all the other members of the colony as they would share with their own offspring.

Among the insects, several groups have evolved caste systems involving reproductive males and both reproductive and nonreproductive females. In these groups, which include the termites, ants, and social bees and wasps, the nonreproductive females have given up the opportunity of producing their own offspring in order to increase the reproductive output of the colony. Some bacteria kill their hosts, thereby committing suicide. In doing so, they make the resources in their host available to other hosts that carry identical copies of themselves. In all these cases, individual reproduction is sacrificed to aid kin that carry the same or a high proportion of the same genes. Examples of this type, in which individual reproduction is sacrificed specifically to increase the transmission of genes as a result of the reproduction of close relatives, which will carry many of the same genes, is termed *kin selection* (p. 144). These examples of kin selection suggest that it may be more accurate to say that the ultimate aim of an individual is to promote the passage of its genes through time whether this is by reproducing itself or by aiding the offspring of others that carry its genes.

Asexual Reproduction

The earliest life on Earth, dating back some 3.8 billion years, or maybe more, undoubtedly reproduced simply by making copies of itself. Many primitive organisms—viruses, bacteria, single-celled algae, and some animals—still reproduce in this way: amoebae, sponges, and sea anemones divide; hydra bud. Many plants, including higher plants have exclusive or partial asexual reproduction through the vegetative production of stolons, rhizomes, leaflets, and tillers. One major problem with asexual reproduction is that all individuals derived from one ancestor would be essentially the same. Only minor variations resulting from rare genetic mutations would exist. Populations of such genetically similar individuals—that by their very existence would have to be well adapted to the conditions in which they lived—would be in danger of extinction, due to the lack of genetic variation, should their environment change. Later, much later, perhaps about 1.5 billion years ago, an alternative form of reproduction arose, which allowed some exchange of genetic material between similar individuals. This involved the temporary fusion of two individuals in such a way that genetic material could reciprocally migrate from one to the other before reproduction. This was the start of sex.

Sexual Reproduction

Initially, sex involved the fusion of whole individuals that were approximately the same size. The individuals in question were small, comprising just a single cell. Considering current single-celled organisms that reproduce in this way, such as the simple algae of the genus *Chlamydomonas*, it appears that at some stage in the early evolution of sexual reproduction, differences in the surface chemistry of individuals arose, whereby only individuals with different surface chemistries could fuse.

Genes with harmful effects are often recessive, which is to say they are only expressed when both copies of a gene in an individual are the same. When closely related individuals mate, the chance that both carry the same harmful version of a gene is increased. The offspring of reproduction between close relatives frequently show deleterious characteristics due to the expression of recessive genes. The reduction in fitness of such individuals compared to the norm is called *inbreeding depression.*

The evolution of differences in the surface chemistries in early sexual reproducers would have provided a mechanism that reduced the possibility of fusing with a genetically very similar or identical relative. The evolution of differences in surface chemistry would thus have been selectively favored because it would reduce the instances of rare, harmful, recessive mutations being inherited from both parental cells. In essence, fusing with unrelated cells avoids inbreeding depression.

In single-celled organisms that reproduce in this way, the cells that fuse are of the same size. In most species, only two types of individual occur with respect to their surface chemistry. These are called mating types, one being designated the + or recipient type, and the other the − or donor type.

But, if the reason for evolving different mating types is to avoid inbreeding, why are there just two mating types? With just two mating types, the potential mates available for each individual are reduced by half, assuming there are 50% of each type. This cut would be avoided if there were many mating types. The answer appears to involve the inheritance of organelles, the organs of cells, such as mitochondria and the photosynthesizing chloroplasts. These are inherited in the cytoplasm of cells, not in the nucleus. These organelles are normally inherited just from the female parent (although there are exceptions; e.g., the chloroplasts of conifers are paternally inherited).

The reason for the almost universal uniparental inheritance of organelles is that if they were inherited from both parents, there would be scope for intracellular warfare. Imagine that mitochondria, the energy-producing organelles of cells, were inherited from both parental cells. Mitochondria, of which there are many copies in each cell, pass randomly into daughter cells when the cell splits. There is no mechanism to ensure that the daughter cells each receive the same number of mitochondria. Therefore, if a strain of mitochondria evolved faster replication, it would spread through the population, even if this selfish behavior were at the expense of its primary function, the production of adenosine triphosphate (ATP), a major carrier of energy in biological systems. The result would be the evolution of more and more selfish, but less efficient mitochondria, to the ultimate detriment of their host cells and the selfish mitochondria themselves.

Confirmation of this theoretical explanation of the limit to just two mating types in almost all species, was synthesized from some remarkable observations of ciliate protozoa, such as *Paramecium* (Hurst and Hamilton 1992). In these ciliates, cells do not fuse. Rather, when two cells lie together, each divides its nucleus. Thus, each forms two haploid nuclei, one of which is passed on to the other cell. Mitochondria are not transferred. Following this exchange of nuclear material, the two cells separate, and the two haploid nuclei that each contains fuse to form a diploid. By this mechanism, nuclear genes originate from two parents, while organelles come from just one. And these ciliates, in which there is no chance of warfare between organelles, have multiple mating types. However, one group of ciliates, the hypotrichs, employs both conjugation (cellular fusion) and gamete fusion. Amazingly, these have multiple mating types for conjugation, but just two for gamete fusion, elegantly confirming the theory.

Sexual reproduction appears to have been successful from the moment it arose. The evolutionary reasons why this novel type of reproduction was successful in competition with asexual reproduction need not concern us for the moment. This complex subject will be discussed in chapter 9.

Sex Cells of Different Sizes

The next major step in the evolution of reproduction came with the rise of multicellular organisms. Once an individual was composed of many cells, the different cells that comprised the whole could be specialized for particular functions. The cells that took the role of fusing together to give rise to new individuals are variously called sex cells, germ cells, gametes, spermatozoa, sperm, oocytes, or eggs. In early multicellular organisms that reproduced sexually, the cells that fused during sexual reproduction were of the same size and may be considered as donor and recipient types. However, in time, the sizes of the donor and recipient cells began to change, the donor mating-type germ cells becoming smaller and the recipient cells larger. Species in which the gametes produced are of the same size are called isogamous. When gametes produced are of different sizes, as occurs in all sexually reproducing higher animals and plants, species are said to be anisogamous.

Geoff Parker and his colleagues (Parker et al. 1972) argued that the change from isogamy to anisogamy was the result of an arms race within a species. They envision a mutation in a population of isogamous individuals that caused one individual to produce slightly smaller germ cells or gametes. The reduction in the cost of production of these smaller gametes would allow this individual to produce greater numbers of gametes. As long as the majority of the rest of the population were producing normal (i.e., larger) gametes, the mutant would be beneficial, as it would have greater reproductive success through the production of more gametes. Thus, it would start to spread through the population. However, as the smaller gamete–producing mutation increased in frequency in the population, the possibility of two of these smaller gametes fusing would increase. There would then be the possibility that the individual produced by the fusion of two small gametes had insufficient nutrient resources to develop successfully. In this case there would be selection in favor of small gametes that only fertilize large gametes. Furthermore, once there was competition between small gamete–producing individuals, there would be a pressure to produce even smaller gametes to increase the number produced. As the small gametes became ever smaller, selection would have imposed pressure on the individuals producing large gametes to produce still larger gametes with more nutrient resources to compensate.

Parker's theory of the evolution of anisogamy was widely accepted. However, other theories have also been proposed. For example, Laurence Hurst (1990) has suggested that inheriting cytoplasm from just one parent may have evolved to reduce the chance of inheriting parasitic organisms that live in the cytoplasm of host cells. The argument is simple, elegant, and self-evident because it is obviously more likely that progeny will inherit a parasite if this can come from both parents rather than just one.

A further idea, proposed by Laurence Hurst and Bill Hamilton (1992), suggests that the driving force behind the evolution of the uniparental inheritance of organelles in the cytoplasm that arises out of anisogamy is the minimization of the potential for conflict between the cytoplasmic genomes of fusing cells. In the same way that two individuals of a species may be in conflict with one another and compete over food, or mates, or some other limited resource, so the cytoplasms that come together at fertilization may compete. The conflict produced will certainly be to the detriment of the individual created. Should nuclear genes that suppress cytoplasmic mixing, by promoting inheritance of cytoplasm from just one parental cell, arise by mutation, such mutations would be favored.

As yet we have insufficient data on the interactions between intracellular parasites and their hosts, or on the conflict between cytoplasmic genomes, with which to judge the validity of these ideas. It may be noted, however, that these three theories for the evolution of anisogamy are not mutually exclusive, and all three may have had an influence.

We may perhaps combine the three to some extent in a summary of the evolution of anisogamy. Here we may consider the relative contributions of the two types of cell to the new individual. Reduction in the size of gametes produced may initially be beneficial, because more gametes can be produced for a set energetic cost. To avoid cytoplasmic conflict, the genes within the cytoplasm carried on some of the organelles in the cytoplasm should only be contributed by one parent. The gametes of this parent thus have to be large enough to contain sufficient organelles. This gamete will also contain the nutrients necessary to fuel the zygote when it is formed. The other parent must then be stripped of its organelles, or the cytoplasm that carries them, so that it contributes just its nucleus and has no requirement to contribute nutrient resources to the zygote formed. The result is that one gamete will be larger than the other. It then makes energetic sense if the smaller gamete remains mobile while the larger gamete loses its mobility.

Males and Females: The Root of Sexual Conflict

Once gametes are of differing sizes, we can refer to them as different sexes, rather than mating types, for the fundamental distinction between males and females depends purely on gamete size. Females produce the larger and males the smaller gamete. No other difference between males and females is universally definitive. Usually, because the small gamete is mobile, the male gamete, or sperm, moves toward the more sedentary female gamete, or egg. However, there are exceptions to this course of action. For instance, in many seahorses and pipe-fish, the eggs are transferred from the female into the male through a penis-shaped organ, and it is the male that becomes pregnant, caring for the fertilized eggs in a specialized pouch.

Sexual Selection: Competing Males and Choosy Females

The evolution of different gamete sizes, the sexes, and the different roles played by the sexes allowed selection to act on the sexes in different ways. The differences in selection are based, at least in part, on the differences in the gametes that males and females produce. Thus, males produce huge numbers of energetically cheap sperm, while females produce far fewer, but much more costly eggs. Males thus have gametes to spare and have a much greater reproductive potential than females, as shown by the maximum number of offspring recorded to have been produced by males and females of a variety of species (Table 1.1). Males can gain by mating repeatedly, which in turn can lead to advantages in having weaponry to compete with other males for mating opportunities, or ornaments that are attractive to females. So males commonly will compete with each other for access to females. Females, on the other hand, should protect the large investment they have put into each of their eggs by doing all they can to ensure that this investment is not devalued through fertilization by sperm with poor genes. Thus, females should evolve mechanisms to assess the genetic quality of male suitors and opt to mate with the male with the best genes available. Females may obtain males of high genetic fitness by acceding sexually to the victors of male-male conflicts, or by selecting mates on the basis of specific traits indicative of high genetic quality. Any heritable character that enhances the probability of a male obtaining a mate, or which allows females to assess male quality, may become the subject of sexual selection.

These two mechanisms—male competition, which Darwin (1859) called the "Law of Battle," and female choice—involve selection that does not neces-

TABLE 1.1
The difference in the reproductive potential of males and females measured as the maximum number of offspring recorded to be produced by males and females of a number of animal species. Man, red deer, elephant seals, two-spot ladybirds, and angleshades moths are polygynous. Kittiwakes are monogamous. (Data from Krebs and Davies 1987; Majerus unpublished.)

	Maximum Number of Offspring Produced During Lifetime	
Species	*Male*	*Female*
Man	888	69
Red deer	24	14
Elephant seal	100	8
Kittiwake	26	28
Two-spot ladybird*	42,415	2,341
Angleshades moth*	17,549	487

*Measured in captivity.

sarily increase an individual's chance of survival, but increases an individual's chances of mating and producing fit offspring. Darwin termed this type of selection *sexual selection* and saw its existence as a prerequisite to explain some of the extreme differences seen between males and females within species.

Darwin cited many examples that are still often quoted in the context of male competition. These include the weapons, such as antlers, horns, and tusks used by male mammals in fights over females; the huge size of males compared to females in elephant seals and some primates; and the array of challenging songs used when rival male birds encounter one another in the breeding season. For female choice, he argued that male traits such as the extravagant plumage and displays of peacocks, male birds of paradise, and lyre birds could be of no direct survival advantage to the males themselves, but must serve to excite females so that they accept the most brightly adorned males.

Weaponry and elaborate adornments are not restricted to the vertebrates. Males and females of many species of insect are strikingly different. Male stag beetles and rhinoceros beetles sport impressive antlers or horns with which to fight other males. Male fig wasps have a variety of aggressive and defensive weapons, including body armor, femoral spikes, and large scythe-like jaws, all of which are used in often fatal battles with other males. Male butterflies are frequently far more brightly colored than their female counterparts, particularly in the ultraviolet spectrum. Some groups of flies have developed what are called stalk eyes, the eyes being set on long, protruding head appendages somewhat reminiscent of the heads of hammerhead sharks. In all these cases and many others, there is evidence that the traits have evolved through sexual selection.

Pre- and Postcopulatory Male Competition

Male competition, or what should more correctly be called *intrasexual selection*, because in a minority of species the roles of the sexes are reversed and females compete with each other while males choose (see p. 171), has had a fairly clear passage since it was first proposed by Darwin (1859) in *The Origin of Species*. It was rapidly accepted as a reasonable thesis and has changed little in emphasis. Perhaps the greatest difference between Darwin's and our contemporary views of male competition is that, rather than thinking purely of conflicts between males to gain mating opportunities with females, we now extend the competition to the point when sperm fertilizes the egg. Thus, male competition now encompasses sperm competition (defined as the competition between sperm from two or more males, within a single female, for fertilization of the eggs) and a range of other devices employed by males to secure paternity once they have copulated by preventing females from mating with other males.

The range of sperm competition strategies has been increasing steadily since Geoff Parker's pioneering paper on the subject (Parker 1970). Examples of elaborations of male genitalia, which allow males to scoop out or displace sperm from previous males, abound in many taxa, particularly insects. One species of crab has the ability to move sperm from previous males to a position within the female where it can be sealed off from the duct that leads to the eggs by a hard-setting gel secreted by the copulating male (Diesel 1990). Some fruit flies deposit into females, via the ejaculate, chemicals that kill or disable stored sperm from previous matings, but leave their own sperm unharmed (Harshman and Prout 1994)

In many species, sperm competition appears less sophisticated, involving little more than a numbers game. If the likelihood of each sperm within a female fertilizing an egg is the same, then just as buying more tickets for a lottery will increase your chance of winning, the male that puts most sperm into a female will father the most offspring from that female. Comparative analysis of levels of multiple mating by females (polygyny) and numbers of sperm in an ejaculate, in a variety of taxa, lend circumstantial weight to this lottery mechanism of sperm competition. For example, in primates, there is a strong correlation between degree of polygyny and male testis and ejaculate size. In male gorillas, where one single male, the silver back, mates with all the female members of his troop so that females only mate with a single male in a reproductive cycle, testis and ejaculate sizes are much smaller than in chimpanzees. In chimps, females may mate with several or, indeed, all of the mature males in their troop. Humans are intermediate between gorillas and chimps with respect to both the mean levels of polygyny and male ejaculate size.

More direct evidence comes from recent studies of an Australian dung beetle. The males are of two types. Most males have "horns" extending in front of them. The horns of these so-called "guard males" are used in competitive fights with other males. Successful guard males pair with females, stay with them, help them to build burrows and resource these with a store of animal dung in which the female will lay an egg. Males of the second type are called "sneaks." These males are smaller, lack horns, and neither fight with other males nor form pair bonds with females, so they do not help females make or provision their egg-laying burrows. Rather, they find digging pairs of beetles and dig another burrow to intersect with the pair's burrow. Then, while the guard male is collecting dung, the sneak will move from his burrow into the main burrow and mate with the female. Analysis of testis size and sperm content of guard and sneak males has shown that the sneaks produce much larger ejaculates. This makes sense because, while many guards will not be sneaked upon and so will be involved in monogamous partnerships, all sneaks will be involved in polygamous pairings. Success for sneaks thus depends specifically on the ability of their sperm to outcompete sperm from the guard males of the females that they mate with. Here then, there is a tradeoff within a species between precopulatory male competition and postcopulatory male competition in the form of sperm competition. Guard males put resources into size, strength, and weaponry to help secure a partner and into parental care through burrow building and provisioning. They thus have fewer resources available for sperm production than do the sneaks who do not expend energy on body building or parental care.

The full extent of mechanisms of sperm competition are as yet unclear, but it is doubtful that we have yet discovered all the means by which males give their sperm a greater chance of achieving fertilization than the sperm of other males. Why, for instance are the spermatozoa of the fruit fly *Drosophila bifurcata* some 58mm long, or 20 times the length of the fly that produces them (Pitnick et al. 1995)? The reason is unknown, but it is difficult to believe that these giant sperm have evolved for reasons other than sperm competition.

Other mechanisms of postcopulatory male competition, which do not fall strictly within the definition of sperm competition, involve strategies that reduce the access of other males to the female a male has just mated with. Thus, males of the South American butterfly *Heliconius erato* inject an anti-aphrodisiac into their mates. This acts to repel other males for several weeks (Gilbert 1976). Tomcats have a less sophisticated method of reducing the likelihood that they will be cuckolded by a queenie soon after he has mated with her. The penis of a tomcat is armed with barbs that inflict considerable damage on the female's vagina as it is withdrawn, thus making further copulation extremely painful for some time afterwards. Many male insects, particularly among the butterflies, the grasshoppers, and the crickets, seal off the copulatory opening of their partners with a plug formed at the end of mating (see, e.g., Ehrlich and Ehrlich 1978). Perhaps the most common postcopulatory mechanism by which males attempt to protect their paternity is the least sophisticated: males just hang onto their females until they have laid eggs. This mate-guarding tactic occurs in many species of insect, some of the best-studied cases involving damsel flies and water striders.

Female Choice

Female choice of males (or again more correctly, *intersexual selection* because of male choice of competing females in sex role–reversed species) has had a much bumpier ride than male competition since it was first proposed by Darwin. Initial resistance probably owed something to chauvinistic attitudes of the Victorian scientific fraternity. More objectively, some critics of female choice, noting that Darwin used the mechanism to explain bright and elaborate adornments, argued that he was imbuing females with a highly developed aesthetic sense: females were choosing the most beautiful males. Furthermore, Darwin, in his writings on the subject, did not propose any explanation of how or why female preferences for particular types of male could have evolved. He simply took their existence to be necessary to explain certain secondary sexual characteristics of males.

Direct and Indirect Benefits to Choosy Females

During the twentieth century, a wide variety of mechanisms for the evolution of female mating preferences have been developed. Females that exercise a preference for particular males may, in so doing, increase their own fitness compared to females that mate randomly. For example, they may receive extra nutrients by choosing to mate with a male that brings them a particularly large piece of food as a nuptial gift, as seen in the hanging fly *Hylobittacus apicalis* (Thornhill 1976). In Clouded Yellow butterflies of the genus *Colias*, females mate preferentially with males sporting bright, immaculate color patterns. The brightness and perfection of the male color patterns is inversely correlated to adult age, which is in turn correlated to the probability of the male having previously mated. Spermatophores in these butterflies contain not only sperm, but also nutrients that are used by females in egg resourcing and maturation. The first spermatophore produced by a male has up to twice the nutrient content of subsequent spermatophores. Females, therefore, gain considerable benefit by choosing to mate with immaculately patterned males that have a high probability of being young and virgin.

Choosy females may make choices using a wide array of criteria. They may pick the males that hold the most resource-rich, or predator-free territories. Benefit may be gained by mating with a strong, fit-looking male rather than one that appears to be in poor health, as this may reduce the chance of contracting parasites from the male during copulation. In all these cases, females gain direct benefits from their mate selection because their own fitness is directly enhanced.

As an alternative to these direct benefits, a female may increase her fitness, not directly in the form of longer personal survival or the production of more progeny, but by increasing the fitness of her progeny. Different indirect benefits theories include Fisher's "sexy sons" hypothesis, Zahavi's "handicap principle," and Hamilton and Zuk's "condition-dependent handicaps."

Sexy Sons

Fisher's sexy sons hypothesis proposes that if a female initially chooses a male because he bears some trait that is an indicator of good genetic quality, this preference will spread because her progeny will be more successful than the norm (Fisher 1930). Female progeny will gain because they are the offspring of a genetically fit father, while sons will gain this advantage plus the advantage of inheriting the trait by which females make their choice. In essence then, the preferred trait and the preference each helps select the other, leading to what has been termed *runaway sexual selection* (Box 1.1). There are several problems with this process. First, as Peter O'Donald (1980) has shown, the process is extremely slow in getting started. If a new preference gene arises in a population of reasonable size, it will initially only be expressed by a single female (we will assume that the gene is genetically dominant). This means that, initially, the only males that gain a reproductive advantage from the existence of this female mating preference are those that mate with this one female. The reproductive advantage arising out of this novel mate preference is thus exceedingly small at first. Fisher (1930) recognized that an evolutionary explanation of both the male trait and the female preference were required, and that the greatest difficulty was in getting the preference started. He solved this problem by proposing two selective influences on the evolution of the male trait:

> i) an initial advantage not due to sexual preference, which advantage may be quite inconsiderable in magnitude, and ii) an additional advantage conferred by female preference, which will be proportional to the intensity of this preference. (p. 136)

Here Fisher is proposing that the initial evolution of female mate choice genes is aided by direct selection. Alternatively, the initial increase in the frequency of mate choice genes would be facilitated if the preference were based on traits associated with male dominance that had already evolved to an extent through male competition.

The second criticism of the Fisherian process is that the maintenance of mating preferences is problematic. Once the female mate preference has spread through the population so that many females are choosing males on the basis of a specific male trait, selection will favor the presence of this trait in all males. Eventually, once all males have the trait, all males will be equal. There will be no genetic variability between males with respect to this trait and so no reason why females should continue to base their choice on it. Only if mutations with respect to the genes controlling the preferred trait tended to cause a reduction in the trait would the preference be maintained (Isawa et al. 1991).

A third difficulty with the sexy sons idea is that it may be prone to invasion by cheats. Males, even if they do not possess good genes for overall fitness, need only produce the chosen trait to gain matings.

Box 1.1 Fisher's "sexy sons" hypothesis of the evolution of sexual selection by female choice.

Initially a novel morphological (or behavioral) trait in males begins to spread due to some natural selective advantage.

A mutation arises that causes females to preferentially mate with males possessing the spreading trait.

Males with the new trait now gain a reproductive advantage, as some females preferentially mate with them rather than with males lacking the natural selective advantage of carrying the trait.

Females that mate with preferred males as a result of the mate preference gene produce fitter progeny, as they are selecting males with a trait that confers a natural selective benefit.

The preference genes and those for the preferred trait thus each increases the advantage of the other, so that they increase together at an ever-increasing rate producing "runaway" sexual selection.

Eventually, development of the preferred trait becomes so extreme, it loses its initial natural selective advantage and becomes detrimental. Further development of the trait due to female choice is then arrested by counter natural selection.

Handicaps and Condition-Dependent Handicaps

Amotz Zahavi (1975) considered this third problem. He argued that an indicator of good genes could only work (i.e., not be prone to cheats) if the preferred trait were costly to produce and could thus only be produced by males that really did have good genes. In consequence, females should only choose on the basis of expensive traits or "handicaps," which would thus be incorruptible indicators of high genetic quality. This hypothesis also has some difficulties because of a lack of a genetic correlation between the costly trait and the good genes. A female that mated with a male bearing the handicap and good genes, may, in addition to producing some progeny that had both the attractive trait (the handicap) and good genes, produce other progeny that lacked one or another of these components of the system. Such males would either have good genes but be rarely chosen by females, or be chosen by females but produce unfit progeny. In either case, they would contribute little to the choosing female's (i.e., their mother's) long-term fitness.

The handicap system only works properly if of those sons carrying the handicap trait genes, only those that are of sufficiently high quality are able to make the trait. Here the development of the preferred trait is dependent on the condition of the males. Sons in poor condition as a result of carrying poor genes do not bear the whole cost of the handicap because they cannot develop it fully. As Bill Hamilton and Marlene Zuk (1982) have suggested, this may occur when there are good genes for parasite resistance. Males with good genes would have fewer parasites, would be in better condition, and so would be able to develop the extravagant display trait.

Sensory Exploitation

A third hypothesis for the evolution of intersexual selection is that mating preferences have evolved as a by-product of pre-existing biases in the sensory abilities of a species. This is the *sensory exploitation hypothesis*. Here, members of the chosen sex take advantage of the likes and dislikes of the choosers. Thus, for example, males may evolve colors that are easily perceived by the visual system of the females, or songs that are in tune with female auditory ability, or scents that impact positively on the olfactory system of potential mates. The traits that males thus develop are the product of sexual selection because males gain a reproductive benefit, not a survival benefit. However, the female preferences are pre-evolved, having developed for some ecological or behavioral reason other than mate choice.

Making use of pre-existing sensory biases that have evolved for other unrelated reasons is easiest to explain by a couple of examples. First, in our own human context, we have a high sensory receptivity to certain colors, such as bright red and bright yellow. The reasons for this receptivity need not concern us, except to say that it was not evolved for use in shopping. Yet commercial companies make use of our receptivity to these colors in the packaging of many of their products.

In a sexual context, recent work on sexual cannibalism in the fishing spider, *Dolomedes fimbriatus*, suggests that the habit of females eating males in this species is not adaptive, but has evolved as a by-product of female feeding behavior that has been selected to increase her fecundity. Arnqvist and Henriksson (1997) have shown that the number of eggs produced by females of this species depends on the female's size in her final immature instar. This, in turn, depends on her willingness to attack prey rapaciously. The more indiscriminate females are in their prey attacks, the more fecund they become. Many males of this species become food for the females that they mate with. Arnqvist and Henriksson showed that this sexual cannibalism benefits neither males nor females. Cannibalistic females do not increase their fecundity by eating their partner. Indeed, because sperm transfer may be interrupted as the male is devoured, fertility is reduced as a result of the cannibalism. Males also seem to gain no benefit. Certainly they show no signs of complicity in the process.

The nonadaptive explanation of sexual cannibalism in this spider is in stark contrast to findings in other spiders. In the garden spider, *Araneus diadematus*, female body mass increases significantly as a result of mate consumption (Elgar and Nash 1988). In the orb-weaving spider, female fecundity increases as well (Sasaki and Iwahashi 1995). In both these cases, it is argued that males gain benefit from their self-sacrifice by increasing the number of progeny produced by the females that they have mated with. In the case of the Australian redback spider, *Latrodectus hasselti*, the male's benefit is more direct. Here, males that are eaten by their mates fertilize more eggs than do those that are not eaten, simply because they stay *in copula* for about twice as long, this additional time being spent by the female in finishing her meal (Andrade 1996).

Mate Choice, Reinforcement, and Speciation

Finally, mating preferences may evolve as a consequence of speciation. Many pairs of species are thought to have evolved as a result of divergence during periods when a single, ancestral population was split by some environmental change or geographic barrier. If, while separated, two species have become sufficiently different genetically that, once they occur together again, hybrids between them are relatively unfit (hybrids have low viability, poor fertility, or are poorly adapted to environmental conditions), they are said to show some degree of postzygotic reproductive isolation. This simply means that the selection against the production of hybrids acts after fertilization between germ cells from two different species has produced a hybrid zygote. In such circumstances, individuals that choose to mate with males of their own species will be selected for, as these animals avoid the energy wastage incurred in producing unfit offspring from hybrid matings. The preference that evolves does not have to be for a trait that is an indicator of mate fitness: it simply has to indicate that the mate is from the same species. Both the preference genes and the genes coding for the preferred trait are likely to spread through a species to fixation because hybrid matings will always be disadvantageous. The result will be the evolution of species-recognition systems that act prior to fertilization and usually prior to mating. Thus, such mating preferences lead to what are termed prezygotic reproductive isolation mechanisms.

This is a reinforcement system, the species-specific mating preference reinforcing the incomplete and wasteful postzygotic reproductive isolation. Such a system could account for the evolution of some differences between males and females of a species, as there is no reason why both males and females have to evolve the preferred trait, although both may do so. Let us assume that males exhibit a species recognition signal and females receive it. As all males should carry the preferred trait and all females in the population should exercise the same mating preference, there does not appear to be much scope here for mate choice within a species. However, there are two possible ways that such female choice could arise. First, at the start of the evolution of a species recognition system, selection will promote any system that allows the correct individuals (i.e., conspecifics) to be chosen, however this preference is expressed behaviorally. Thus, the preference could be for a specific trait, such as a bright color spot of specific dimensions, or for the most expressed trait, perhaps the biggest bright-colored spot. This means that, in some cases, a trait may become extremely exaggerated simply because the species recognition gene(s) that first arose, involved a "supernormal" stimulus. A stimulus is referred to as supernormal if the extent of the response to the stimulus is correlated to the size of the stimulus, even if the size of the stimulus is larger than that which the receiver normally encounters. If long tails, for instance, indicate to females of one of the two diverging species that males were from their own species, then the longer the tail of a male that one of these females mated with, the more positive she would be that she was making the right choice.

Alternatively, intraspecific mating preferences could evolve later as a consequence of the existence of species-recognition genes if the selective pressures that originally led to their evolution were relaxed. Should, for example, one of the two species become extinct or move away so the two species no longer came into contact, the selective constraints on the recognition system would be removed. The genes involved would then be free to evolve in new directions, one possible direction being that they would come to act as intraspecific mate fitness indicators.

There are pieces of evidence from a wide variety of examples to support all the preceding theories of the evolution of female choice. However, most of the evidence is circumstantial and much of it does not fully distinguish between alternative theories. In all probability, female mate choice has evolved as a result of each of these processes in some species. Indeed, as many of the theories are not mutually exclusive, two or more of the selection scenarios detailed above may have been involved in some cases.

Female choice has been a hot topic for research over the last thirty years. In particular, the development of molecular tools for the analysis of paternity in the 1980s has allowed for studies of mate choice with much greater rigor and stringency than was previously possible. Few evolutionary geneticists would now argue that females are passive partners in mating. Females of many many species do choose their mates and their choices have a genetic basis. Much work is currently in progress in this field. However, as with male competition, it is now realized that female choice may operate not only prior to copulation, but also during and after copulation.

Cryptic Female Choice

In species in which females mate with more than one male and have the ability to store sperm, there is the opportunity not only for sperm competition, but also for females to have an influence over which sperm succeed in fertilization. This type of postcopulatory female choice has been termed *cryptic female choice*. William Eberhard (1996) argues that females have much greater postcopulatory potential to influence paternity than do males. He produces an impressive list of the ways that females may influence the success of a particular copulation. This list, split into those mechanisms for which cases are known, and those as yet unknown to occur in any species, is (with minor editorial changes) provided in Box 1.2.

Work investigating the role of cryptic female choice is already providing experimental and field data suggesting that females have a considerable degree of control and that in some cases the control is highly sophisticated. To give just a couple of recent examples, Tom Pizzari and Tim Birkhead (2000) have shown that feral female chickens eject sperm of subdominant males. Emma Cunningham and Andrew Russell (2000), working on mallard ducks, have shown that females alter the resources they put into eggs depending on the attractiveness of the drake that they mate with. Females that mate with attractive drakes in good condition produce heavier eggs. Similar results have also been found with zebra finches, *Taeniopygia guttata*, but here, in addition, females deposit into eggs more androgens, such as testosterone, if they have mated with attractive males. These androgens cause chicks in the nest to beg more and to grow faster and results in a higher social rank once fledged.

The yellow dung fly, *Scathophaga stercornaria*, has become a model species in which to study sperm competition and cryptic female choice. The female sperm storage organs of these flies consist of a bursa copulatrix and several spermathecae (Figure 1.1). Research, using radioactively labeled sperm, has shown that females have a considerable influence on the movement of sperm between the bursa copulatrix and the spermathecae (Simmons et al. 1996). Furthermore, it appears that females have control over which storage organ they release sperm from to fertilize their eggs. In an elegant field study, Paul Ward (1998) examined flies that developed from eggs collected from sunny or shaded portions of cow-pats. He found that the flies differed with respect to an enzyme, phosphoglucomustase, which affects larval growth at different temperatures. The different forms of this enzyme are controlled by different alleles of a gene.

Box 1.2 William Eberhard's listing of female behaviors that are known to or that may potentially influence the success of a particular copulation. (Based on Eberhard 1996, with minor editorial changes.) (See Eberhard 1996 for examples.)

Mechanisms for Which Cases are Known

Premature interruption of copulation

Denial of deeper penetration of the male genitalia to internal sites where the male's sperm will have a better chance of being used

Lack of sperm transport to storage and/or fertilization sites within the female

Discharge or digestion of the current male's sperm or those of previous or subsequent males

Lack of ovulation

Lack of preparation of the uterus for implantation of embryos

Abortion

Lack of oviposition

Rejection or removal of mating plugs

Prevention of removal of mating plugs by subsequent males

Removal of sperm-injecting structures (spermatophores) before their contents have been transferred

Selective use of stored sperm

Failure to trigger sperm-injecting mechanisms of spermatophores

Failure to modify insemination ducts, making remating more difficult

Selective fusion with sperm that have reached the egg

Failure to fully resource a particular male's offspring

Lack of rejection of subsequent advances by other males

Potential Mechanisms, as yet Not Observed in any Species

Lack of sperm activation

Lack of sperm nourishment

Failure to seize and open spermatophores in the female's bursa copulatrix

Changes in sperm leakage

Changes in the efficiency of sperm usage

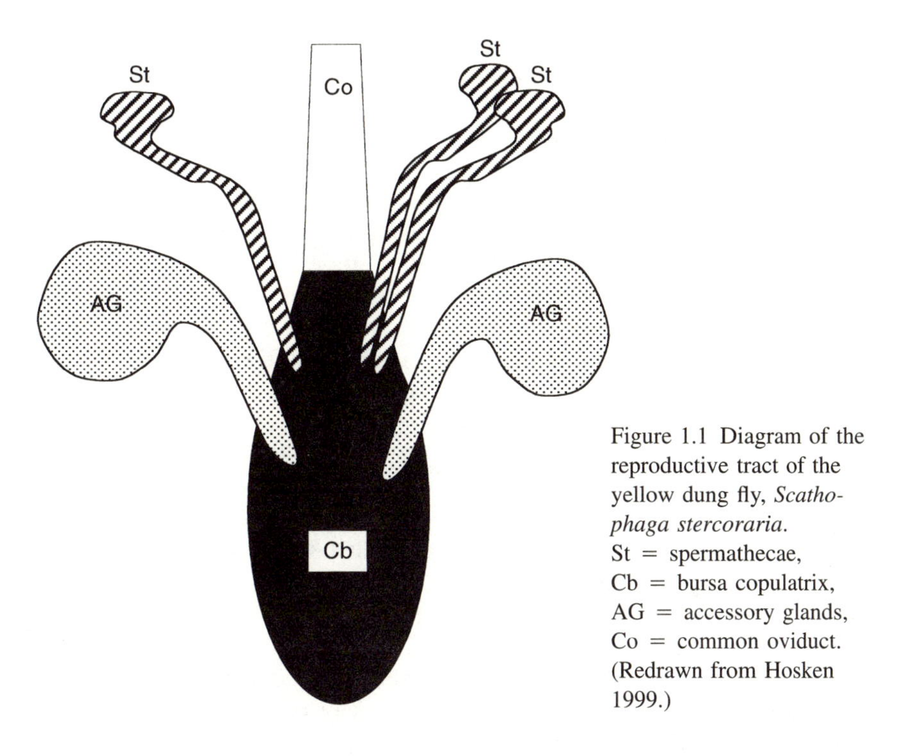

Figure 1.1 Diagram of the reproductive tract of the yellow dung fly, *Scathophaga stercoraria*. St = spermathecae, Cb = bursa copulatrix, AG = accessory glands, Co = common oviduct. (Redrawn from Hosken 1999.)

Ward's results suggest that, as fertilization occurs as eggs are laid, females selectively use sperm with particular alleles for this enzyme depending on their oviposition site on a cow-pat, that is, whether they were laying eggs in sunny or shaded positions.

Teasing out the intricacies involved in sexual selection in any species is not easy. Perhaps the greatest difficulty in the past has been in determining whether mating biases have been due to male competition or female choice. With the additional focuses of sperm competition and cryptic female choice, this task has not become simpler, although we now have powerful molecular genetic tools to help unravel the intricate interactions between males and females that may be involved in sexual selection.

Three final points are worth making in this context. First, with respect to female choice, almost all attention has been focused on the benefits that accrue to females that actively choose their mates or only accede to victorious males. Very little attention has been paid to the costs of female choice. However, there is a small but growing body of evidence to suggest that every time a female mates or rejects an unsuitable male, some cost is incurred. Just think of the wastage of time and energy employed in selecting or rejecting males, or the chances of contracting sexually transmitted diseases during each sexual encounter, or the possibility of injury to passive female bystanders as huge male elephant seals battle for their favors.

In an elegant experiment, Holland and Rice (1999) recently focused on the costs to mating for females. They imposed monogamy on naturally promiscuous fruit flies, thereby all but eradicating sexual selection from their laboratory fly cultures. Their reasoning was that if sexual selection imposed a cost on females, the removal of such selection should result, after a number of generations, in male flies that caused less harm to females. This prediction was fulfilled. After 34 generations of imposed monogamy, both the reproductive rate and the longevity of female flies mated to males from the monogamous cultures were greater than for females mated to control promiscuous flies.

Here then is a further dimension to sexual selection that requires attention. If mating is costly for females, species in which females mate more often than necessary to maintain high fertility require explanation. Of course, it is possible that excess matings may act as an insurance policy against the chance that the first male is infertile. However, there is a growing body of evidence in invertebrates, such as spiders and grasshoppers (see Pardo et al. 1995; Schneider and Elgar 1998) that polyandry increases the genetic variability of progeny. In variable environments, this production of greater progenic variability may be a bet-hedging strategy, which is selectively beneficial because at least some of a female's offspring will have appropriate genes whatever the environment (Yasui 1998).

Second, although in the discussion above, and in most texts on sexual selection, male competition and female choice are treated separately, this does not mean that they are mutually exclusive. Both may operate together, as may sperm competition and cryptic female choice. An example often cited involves elephant seals. Here "beach masters," winners in violent battles with other males, collect, protect, and mate with harems of females. Males that have not secured a harem will attempt to sneak into a harem and mate with a female when the attention of the beach master is elsewhere. The very much smaller females have little chance of physically resisting the interloper. However, it is not in her interests to mate with a male whose genetic credentials have not been proven in the battle for the beach. Consequently, she exercises her choice in the only way she can: she protests loudly, screaming to attract the attention of her beach master, rousing him to attack and drive away the interloper.

The interaction between male competition and female choice is perhaps illustrated even more potently by the reproductive antics of a group of marine flatworms. These flatworms are hermaphrodite and their copulatory behavior shows not only that male competition and female choice go hand in hand, but also beautifully illustrates the difference in the energetic costs of male and female gamete production. These beautifully colored organisms, which come in a dazzling array of colors and patterns, compete with each other to be "male" when two of the same species meet. Both have the capacity to contribute either eggs or sperm to the partnership. However, there is no common agreement as to which partner is to contribute sperm and which eggs. Both jockey for position and form an intromittent organ so that they can inject sperm into the other. The first to succeed is effectively the male, as sperm are only transferred one way. The rationale behind this competition is that sperm are much cheaper to produce than eggs, so the sperm donor apparently gets the better end of the deal. The "female" in this copulation is not a complete loser, however, for although she has lost the battle to be "male," the victor has proved himself by "his" victory over "her," so she gains sperm carrying good genes from her mate.

Third, theoretical and experimental scientists have recently begun to turn their attention to the evolutionary causes of extravagant traits in females. Females of many species of both vertebrate and invertebrate possess showy traits. Many female birds, such as grebes, auks, and shags, have conspicuous crests, lappets, and bills. Bright plumage occurs in female hummingbirds, toucans, and parrots as well as in the males. Some mammals have horns or tusks or antlers in both sexes. Many fish and marine invertebrates are brightly colored in both sexes. In diurnal Lepidoptera, bright ultraviolet color patterns are not always confined to males. Some of these traits may result from the lack of genetic systems that completely limit to males the expression of secondary sexual characters that are beneficial to males for fighting or increasing allure to females. Others may be a consequence of accurate species recognition being beneficial to both sexes. However, there is increasing theoretical and empirical evidence suggesting that some of these traits are the result of either competition between females or male mate choice or both (see Johnstone et al. 1996, Amundsen 2000).

The complexities of reproductive systems and conflicts of interests between males and females, resulting, at least in part, from the disparity in size of male and female gametes, has resulted in a truly extraordinary array of variations on the theme of sexual reproduction. Almost every conceivable way of having sex (and many that, until they were observed, were probably beyond conception) occurs in some species or other. So it is perhaps not surprising that some species have reverted to an asexual lifestyle.

Secondary Asexual Reproduction: Parthenogenesis

One major hazard for sexually reproducing organisms is the need to find and secure a mate. For males, this need is at the root of male competition. Here securing a mate is often more problematic than finding a mate. This is taken to extremes in some spiders and a few insects, in which males not only have to compete with each other for access to females, but also have to give the right signals to females or they end up as the female's lunch rather than her mating partner. For females of most species, securing a mate once found is rarely a problem. However, finding a mate at the right time, that is, when the female is receptive, may be problematic, particularly in rare species scattered at low density. Many mechanisms for attracting males have evolved in females. Scent, in particular, plays a crucial role in many animals. In some silk moths, for example, males may by attracted across 10km or more by a "calling" virgin female. The attraction is very efficient, as evidenced when single female Emperor moth,

Saturnia pavonia (Plate 1a), attracted over 40 males to the cage she was hanging in, within half an hour of starting to call.

But what happens if a female fails to find a mate when she needs one? A female mammal that does not obtain sperm to fertilize her eggs will simply die without issue. But some invertebrates have another option, for in some species, eggs may develop without being fertilized. This type of reproduction is known as *parthenogenesis*. As these species have evolved from sexually reproducing ancestors, the phenomenon is sometimes also referred to as *secondary asexual reproduction*.

Several types of parthenogenesis are known and they have led to two different classifications of parthenogenetic organisms, one based on the genetic mechanism of parthenogenesis, the other on the sex that the unfertilized egg develops into. These two classifications are given in Box 1.3.

In the genetically based classification, the important difference between the classes is whether, as in the case of apomixis and endomitosis, the offspring are genetically identical to their parent (clones), or as in the case of automixis, different from it in that they have increased homozygosity. Given that increased homozygosity is generally considered disadvantageous because it allows the expression of more deleterious recessive genes (p. 32), it is perhaps not surprising that automixis is rather rare. However, there is no such problem with apomixis or endomitosis and we must ask why these forms of reproduction are not commoner, because not only is the need to find a mate avoided, but the cost of producing needless males is avoided.

The Twofold Cost of Sex

The evolutionary argument that leads to describing males as needless may require some explanation. If we imagine an organism that reproduces sexually, has no parental care further than provisioning the eggs, and produces equal numbers of males and females, we might think of a cockroach. Then let a mutation arise in a female cockroach. This mutation achieves two outcomes. First, it suppresses meiosis, the specialized cell division that gives rise to sex cells with a single set of chromosomes (haploid) from cells with two sets of chromosomes (diploid), so that diploid eggs are formed. Second, it causes these eggs to develop into females with the same genetic makeup as their parent. Theoretically, such a gene would be bound to spread. This is easy to see if the situation is considered from the gene's point of view. In the asexual female, all the offspring would get a copy of the meiosis-suppressing mutation and would be female. However, only half of the offspring of a normal female, that is, one that still produces sexually, would gain any particular gene she bore, say, a gene that caused meiosis, because her genes have to be combined equally with male genes in the offspring (Figure 1.2). The result is that when still rare, a mutation that suppresses meiosis and promotes apomixis (or endomitosis) should double in frequency each generation as a result of the so-called "twofold cost of sex": the production of males.

Box 1.3 Classifications of parthenogenetic organisms.

Classification Based on Genetic Mechanism of Parthenogenetic Reproduction

Haplo-diploidy: Meiosis in females is normal, the chromosomes doubling once, but dividing twice so that haploid oocytes result. Oocytes that are fertilized by sperm develop into females, while unfertilized oocytes remain haploid and develop into males. Haplo-diploidy is characteristic of the Hymenoptera and a number of smaller arthropod groups.

Apomixis: The reduction division of meiosis is suppressed with the result that all offspring have the same genetic constitution as their mother. Apomictic parthenogenesis is common in cockroaches (blattids), sawflies (tenthredinids), weevils (curculionids), and aphids.

Automixis: Normal meiosis leads to haploid products. Meiosis is followed by the fusion of two of the products of meiosis to restore diploidy. Alternatively, two genetically identical nuclei produced at the first mitotic division of the haploid egg division fuse to restore the diploid chromosome number. Exceptionally, in *Solenobia* (Lepidoptera) two pairs of nuclei fuse after the second mitotic division. Automixis produces offspring that are not identical to their parent, for it leads to an increase in homozygosity, some or all of the loci at which the parent was heterozygous becoming homozygous.

Endomitosis: Meiosis is preceded by chromosome doubling so that the cells that undergo meiosis are tetraploid. During meiosis, the newly replicated chromosomes pair with their identical copy and meiosis proceeds normally, giving rise to four diploid products. The resulting progeny are all identical to their parent. Endomitosis occurs in some species of lizard, stick insects (phasmids), scale insects (coccids), and psycids.

Classification Based on the Sex Produced from Unfertilized Eggs

Arrhenotoky: Unfertilized eggs produce only males.

Thelytoky: Unfertilized eggs produce only females.

Amphitoky: Unfertilized eggs may produce either males or females.

The Diversity of Parthenogens

Parthenogenetic species occur in most of the major plant and animal taxa, with the exception of gymnosperms and mammals (p. 189). Yet, most parthenogens are closely related to species that reproduce sexually. Indeed, some species comprise some parthenogenetic and some sexual individuals. For example, bugs of the genus *Lecanium* (Homoptera) have some populations that consist solely

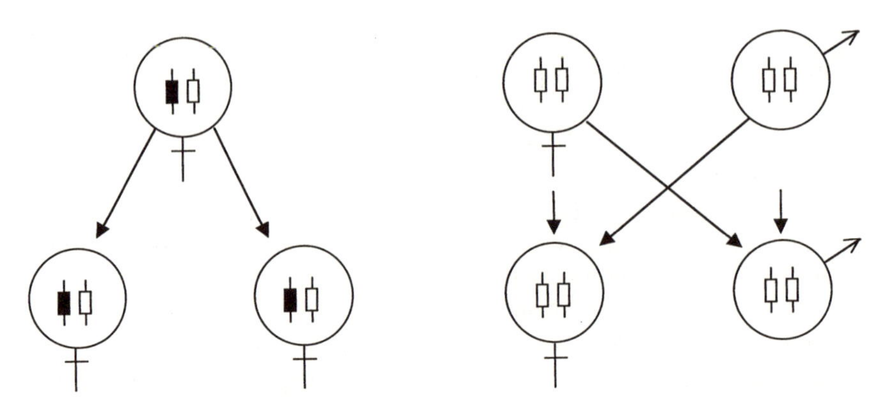

Figure 1.2 The twofold cost of sex. Here the blocked symbol represents a dominant allele that suppresses meiosis, so that the female produces diploid offspring identical to herself. The open allele represents an allele of the same gene that allows meiosis. If females with or without the asexual allele are equally fecund, the frequency of the asexual allele doubles each generation when this allele is rare. (After Maynard Smith 1989.)

of apomictically reproducing females; others consist of males and females with females having the ability to produce female offspring either sexually or by automictic parthenogenesis, with males only resulting from fertilized eggs. In the weevil *Otiorrhynchus dubius*, parthenogenetic populations occur in the north of Europe, while farther south sexual populations occur. This is a common pattern, with the frequency of parthenogenesis being greater at high latitudes. So, for example, in this genus of weevils, 78% of species occurring in Scandinavia reproduce parthenogenetically, while only 28% of those in the Austrian Alps do so.

In the psychid moth, *Dahlica triquetrella*, females are wingless. Examination of the chromosome complements of these females has shown that three types exist in the wild. Some have two sets of 31 chromosomes giving a diploid number of 62 (60 autosomes plus a Z and a W sex chromosome) (ZW). Others have two sets but have only 61 chromosomes because they lacked the W sex chromosome (ZO). The third type had four sets of chromosomes (tetraploid), comprising four of each of the autosomes but just two sex chromosomes, both Z sex chromosomes. Normally, tetraploid females reproduce by thelytokous parthenogenesis. Of the diploids, ZW females reproduce sexually in the normal way while ZO females, like the tetraploids, produce just daughters without mating. In fact, the situation is slightly more complicated than this, because ZO females can reproduce either sexually or parthenogenetically, depending on whether sperm is available for fertilization. In captivity, tetraploid females can be induced to mate with diploid males. Such crosses produce offspring of two types, parthenogenetic tetraploid females, and triploid intersexes with three sets of chromosomes. This shows that the tetraploid females produce diploid gametes that may either be fertilized to produce triploid offspring, or may effectively duplicate without fertilization to produce tetraploids (Seiler 1959, 1960).

An unusual type of parthenogenesis occurs in a "form" of the beetle *Ptinus clavipes*. Normal forms of the beetle are diploid and reproduce sexually. However, the form *nobilis* is triploid and exists only as females. These females produce eggs parthenogenetically, but embryo development is triggered by presence of healthy sperm from normal diploid *P. clavipes*.

Horses for Courses

It is already apparent from the above cases that some insects manage the neat trick of combining both sexual and asexual reproduction. This trick is of course employed by many species of plant that can reproduce sexually through pollen and seeds and can also reproduce vegetatively through stolons, rhizomes, corms, leaflets, and so on.

Perhaps the classic example of this ability in animals involves the aphids. Many species of aphid have an annual cycle that involves an alternation between sexual and asexual reproduction (Figure 1.3). Typically, sexual reproduction occurs in the autumn. Eggs are produced at this time. The eggs overwinter and give rise to asexually reproducing females. These reproduce thelytokously and several parthenogenetic generations follow. Rapid reproduction is achieved at this time because within a female females will be developing that already have developing embryos inside them and, of course, these have no need to spend time finding a mate. Late in the year, sexually reproducing males and females are again produced. The mechanisms by which this transition between sexual and asexual generations is achieved are explained in chapter 3.

Some parthenogenetic insects are also able to reproduce paedogenetically, which is to say that they can reproduce before they have reached the adult stage. This gives rise to some truly remarkable life histories. We have already encountered aphid species in which development of offspring begins in a female before she herself is born, but is still within the reproductive ducts of her mother. However, these young will not be born until the female has reached adulthood. However, in other cases no such requirement need be fulfilled. For example, in beetles of the genus *Micromalthus*, both adults and larvae can reproduce. Males and females are sexual in the normal way. Three types of larvae occur: those that give rise to males, those that give rise to females, and those that produce both male and female progeny. The life cycle is complex (Figure 1.4). Sexual females produce eggs that hatch into triungulin larvae that molt into legless larvae. These legless larvae may feed up and, via a pupal stage, may produce adult females. Alternatively, the legless larvae may produce triungulin larvae that, in turn, change into more legless larvae. A third option is that they may metamorphose into male-producing larvae. These larvae lay a single egg that sticks to the parent larva. When this egg hatches, the new larva eats the parental larva, subsequently pupating and eclosing to become a male. If the egg fails to hatch or becomes detached from its parent, so that the parental

Winter
Eggs on spindle tree.
Eggs have to pass through a cold period before they begin to develop.

Spring
Eggs hatch into females on spindle. These reproduce parthenogenetically. Colonies on spindle increase. In early summer winged females are produced.

Late Spring/Early Summer
Winged parthenogenetic females migrate to secondary host (beans, sugarbeet, dock, poppy, etc.).

Summer
Successive generations of parthenogenetic females. Most lack wings, but as a host plant deteriorates, winged females are produced. These migrate to new host plants, on which new parthenogenetic colonies form.

Early Autumn
Winged females are produced. These fly to the primary host spindle tree where they produce sexual females. Winged males are produced on secondary host at the same time, and these migrate to spindle.

Late Autumn
Sexual females and males mate. Eggs are laid on the stems or buds of spindle trees.

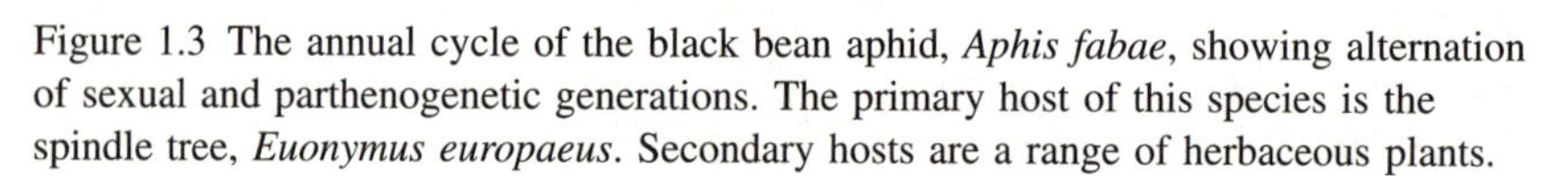

Figure 1.3 The annual cycle of the black bean aphid, *Aphis fabae*, showing alternation of sexual and parthenogenetic generations. The primary host of this species is the spindle tree, *Euonymus europaeus*. Secondary hosts are a range of herbaceous plants.

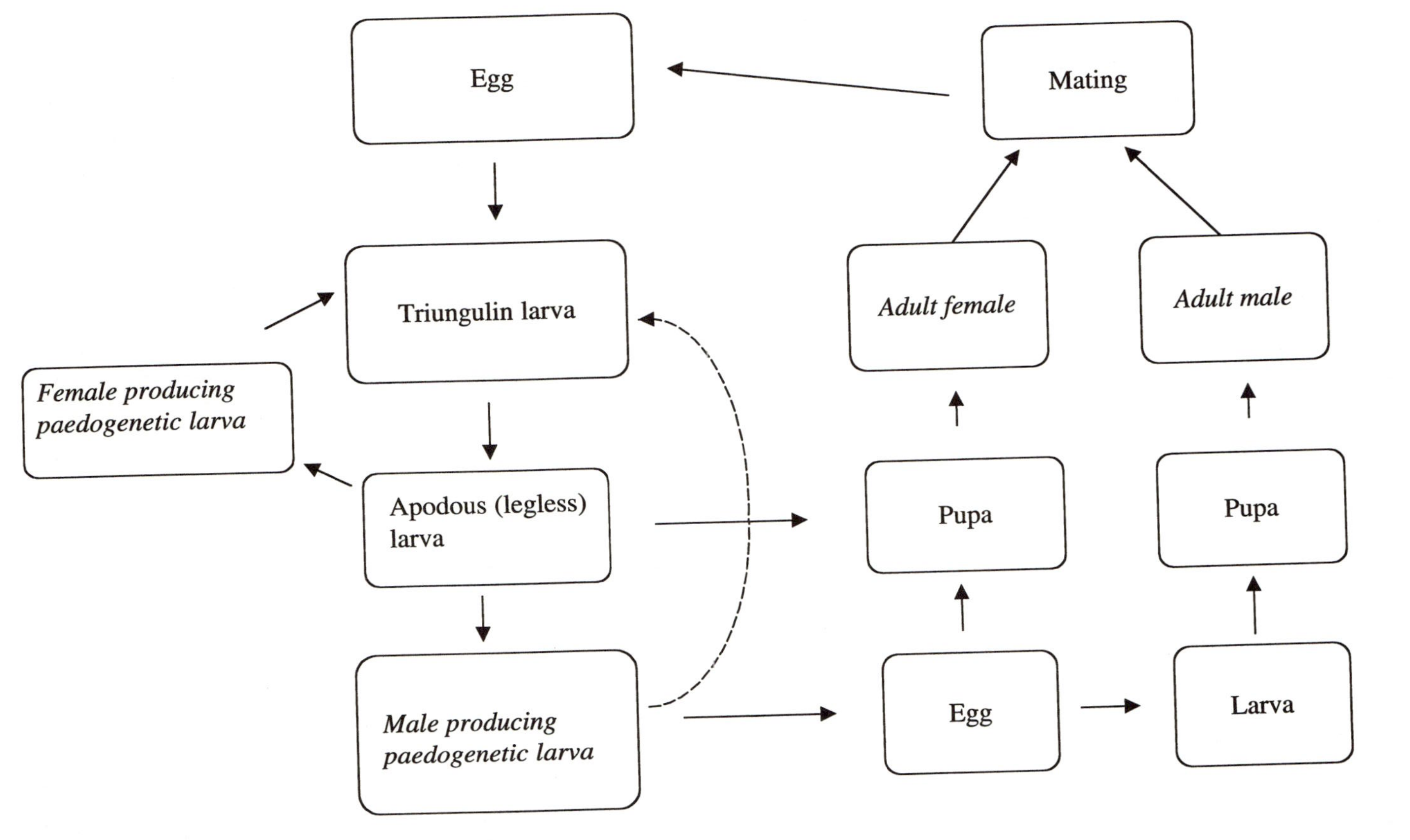

Figure 1.4 The complex life cycle of *Micromalthus*. Reproductive stages are given in italics. (Based on Pringle 1938.)

larva is not consumed, it will subsequently produce a small brood of triungulin larvae.

In the gall midges *Tekomyia* and *Henria* (Cecidomyridae), it is the pupae that give birth. Some larvae of these midges produce pupae and ultimately normal adults. Other larvae metamorphose into a rounded structure called a hemipupa. This hemipupa in turn ruptures to free 30–60 larvae that then develop into pupae.

Another gall midge, *Oligarces* (Cecidomyridae), has taken things a step further. Within the body of parent larvae, unfertilized eggs give rise to daughter larvae that consume and eat their way out of the parent larva. These daughter larvae usually repeat the process and, indeed, paedogenesis is the normal form of reproduction in this species. However, some larvae give rise to male and female adults, although it is not known whether these are capable of reproduction.

Although paedogenesis is usually associated with parthenogenesis, it is interesting to note that a few examples of sexual paedogenesis exist. The bug *Hesperoctenes* is a case in point. Here males mate with larvae by inserting their sipho through the body wall and depositing sperm into the main body cavity. These sperm then migrate to fertilize eggs in the ovaries of the larva and the eggs begin to develop.

Why Is Parthenogenesis Rare?

Despite these fascinating and surprising cases, it is still true that sexual reproduction holds sway in the animal kingdom, so we still have to address the question: why is parthenogenesis so rare, even though it has evolved many times in different evolutionary lineages? Two reasons are usually given and the two are not mutually exclusive. First, it can be shown that parthenogenetic forms evolve more slowly than sexual forms. Second, it has been suggested that parthenogenetic forms will be prone to the accumulation of harmful mutations. These two reasons have wider application because they relate to the general advantages of sexual over asexual reproduction.

Sex Accelerates Evolution

We can begin with a consideration of how sexual and asexual populations may be able to adapt to a changing environment. In an asexual species, rare mutations will give rise to rare individuals with new genotypes, so the population will consist of genetically different individuals, although the novel genotypes will be rare. But, if mutations are occurring at more than one locus, how likely is it that a single individual will carry multiple mutants? Box 1.4 considers the fate of two genes in a haploid, single-celled organism reproducing either asexually or sexually.

The calculation shows that the chance of two advantageous mutations occurring in the same individual is approximately 250 times greater in the sexual

Box 1.4 Sex accelerates evolution.

Let mutations be from A to a and from B to b, each occurring at a rate of 1×10^{-5} (thus a single mutation will occur in 1 in 100,000 cells per cell generation). The chance of both mutations happening simultaneously in a single cell is therefore 1×10^{-10}. Mutations at different gene loci are thus likely to occur in separate cells. But these mutations are recurring and may, after some time, become established at low frequencies in the population. If each mutant has a frequency of 1%, the population will consist of:

Genotype	*Frequency*
AB	98%
Ab	1%
aB	1%

Given the existence of Ab and aB, the double mutant (ab) could arise in a single step. The rate of production would now be:

$$2[10^{-5} \times 10^{-2}] = 2 \times 10^{-7} \text{ (ignoring double mutations)}$$

This is the rate for a population reproducing by asexual means only. We can compare this with a sexually reproducing organism.

In a hypothetical single-cell organism that is haploid (i.e., fertilization is immediately followed by meiosis), with a population consisting of 98% AB, 1% Ab, and 1% aB, the chance of fertilization between Ab and aB individuals is $2 \times (10^{-2})^2 = 2 \times 10^{-4}$.

If such mating occurs, the diploid zygote will be heterozygous for both genes and a quarter of the progeny following meiosis will be ab (assuming no genetic linkage). Thus the probability of producing the ab genotype is:

$$0.25 \times 2 \times 10^{-4} = 5 \times 10^{-5}$$

This is 250 times the chance with asexual reproduction.

population than if reproduction is asexual. Here, then, we have a situation in which, if a population is faced with a changing environment and needs to adapt if it is to survive, sexual reproduction is advantageous over asexual reproduction. This is simply because sex leads to more rapid generation of a wide array of variants, some of which may be able to cope with the new circumstances.

This idea leads to the feeling that when the environment for an individual is suitable and stable, asexual reproduction may be beneficial, but in a harsh or changing environment, sexual reproduction may pay. It is perhaps interesting to reflect that in aphids that show an alteration of generations, the sexual generation generally occurs as conditions deteriorate toward the end of the summer. It has also occurred to me that the advice given by many horticulturist experts with respect to persuading many potted plants to flower has a bearing on this idea of favorable and stressed conditions. For many types of houseplant, keeping the plant pot-bound will cause it to flower, whereas if the plants are regularly transplanted into larger pots, so they have plenty of room to grow vegetatively, this is exactly what they do. The result is a vibrant well-foliated plant with an absence of flowers and sex.

Muller's Ratchet

Several hypotheses have postulated that the advantage of sex lies in the removal of deleterious mutations. One suggests that sex may prevent harmful mutations from accumulating in populations. This theory is known as "Muller's ratchet" after the American geneticist Hermann J. Muller who first proposed it (Muller 1932). The idea is very simple. Most mutations, if they affect the fitness of their carriers at all, are detrimental because most beneficial mutations will already have been selected to fixation in the population. Although mutation in a particular gene is a relatively rare event, there are a lot of genes and so a lot of mutation. In an asexual population without recombination, each lineage will get its share of at least mildly deleterious mutations simply by chance. There will be a tendency, over time, for these slightly deleterious mutations to accumulate because there is no easy mechanism for purging a lineage of its harmful mutants. To see that this is so, we may classify a population according to the number of deleterious mutations carried by individuals: some will have none, some one, some two, and so on. Over time, the number will tend to increase rather than decrease. Now, consider an occasion when the number of individuals with no such mutations is small because population size happens to be at low ebb. There will then be a chance in each generation that, despite their highest fitness, all individuals with no harmful mutations will die without leaving offspring. If so, that optimal class is lost and can only be reconstituted by the rare event of back mutation. In Muller's view, the ratchet has clicked around one notch (Figure 1.5). The new optimal class will carry one deleterious mutation. In time this class may also be lost if the ratchet clicks again. Compare this scenario to that in a sexual population. Here, two individuals, each carrying a different harmful mutation, can, by breeding together, produce some progeny that are free of both mutations. The optimal class is thus continuously reconstituted.

A difficulty with Muller's theory is that while it may function well in small populations, in larger populations there will be little effect, at least in the short term. Kondrashov (1982, 1984) suggested an extension of this idea that would lead to the removal of deleterious mutations in both small and large popula-

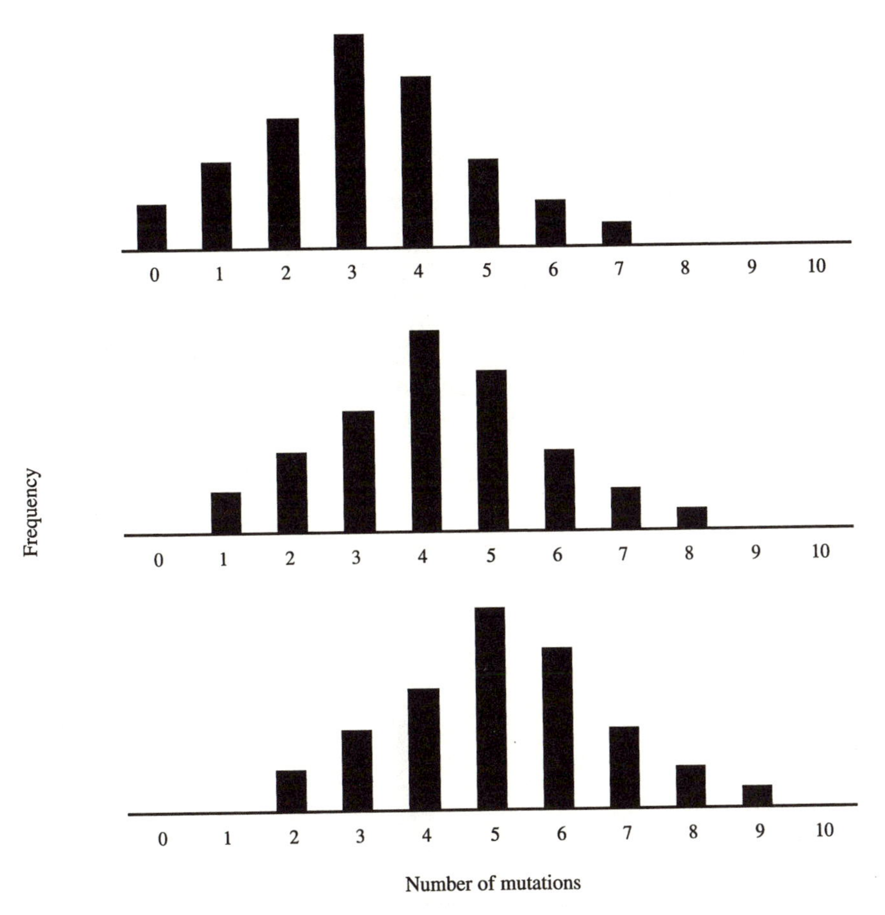

Figure 1.5 Muller's ratchet. The top histogram shows the distribution of individuals with different numbers of slightly deleterious mutations in an asexual population. The middle and bottom histograms show the same population, later in time, when the ratchet has clicked around once and twice, respectively.

tions. Kondrashov's model assumes that the effects of mutations are additive. Imagine two gene loci, with normal alleles *A* and *B* and deleterious mutant alleles *a* and *b*. Normal individuals will be *AB*. Individuals carrying one mutant allele (*Ab* or *aB*) will be less fit, and the double mutant *ab* is lethal. In asexual populations, genotypes such as *Ab* and *aB* may persist for some period in the short term because the lethal genotype *ab* will arise rarely. However, in a sexual population this genotype will regularly be produced if both *a* and *b* are present in the population, and these will thus quickly be purged. Over larger numbers of loci, the variance in the number of deleterious mutations is greater in sexual than in asexual populations. This allows selection to remove many deleterious mutations simultaneously when they occur together (Figure 1.6).

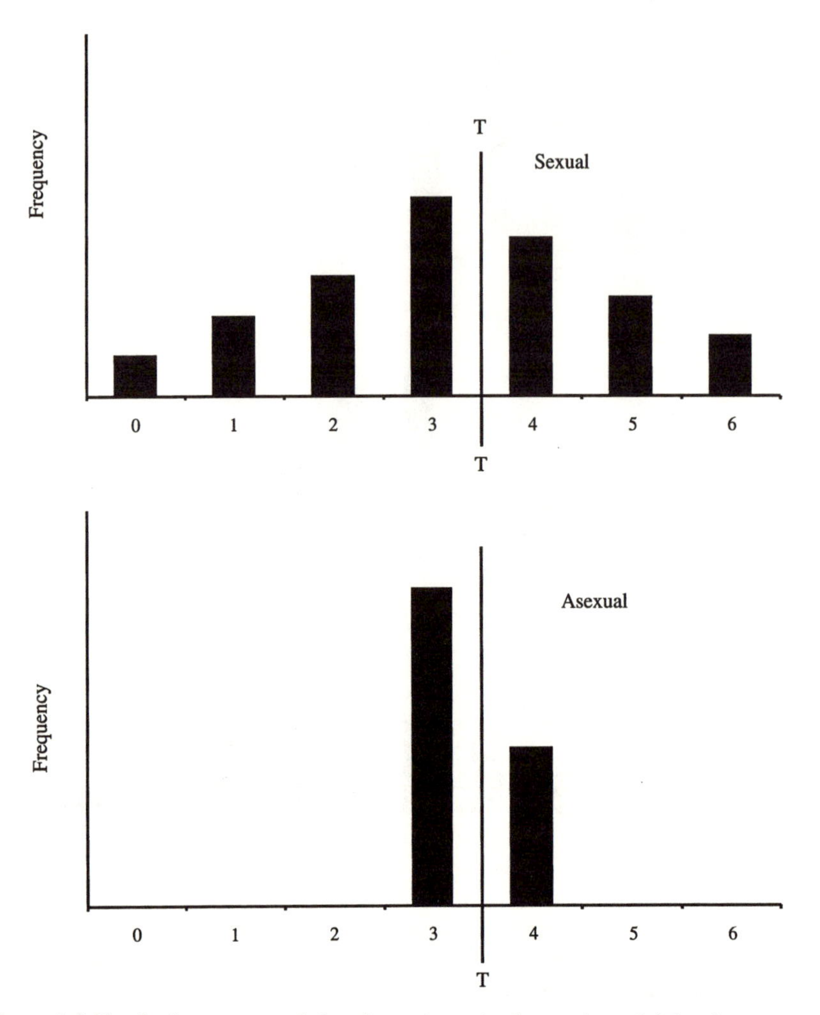

Figure 1.6 Kondrashov suggested that the variance in the number of deleterious mutations per individual would be greater in sexual than in asexual populations. Selective elimination of a fixed proportion of each population with the most deleterious mutations would mean that more harmful mutations would be eliminated from sexual than asexual populations.

The Maintenance of Sex

Both types of hypothesis outlined above involve the fitness of populations. In the first, sex increases the adaptability of the population in a changing environment. In the second, sex prevents a gradual reduction in the fitness of the population by providing a mechanism by which individuals free from harmful mutations can be produced and, due to their high fitness, then be selectively favored. The bottom line is that sexual populations are likely to be longer lived than asexual populations. The advantage to sex here is thus a long-term one (although there may also be short-term benefits in some circumstances, see chapter 9). This is a problem if one is trying to explain the original evolution of sex, for the initial twofold cost of sex would surely mean that, in the short term, asexual individuals would be fitter than those that indulge in sex would be. In considering whether asexual reproduction can evolve from sexual lineages, however, we are looking at things the other way around. Now, the short-term advantage of asexual reproduction can allow for the evolution of asexual lineages from sexual ancestors. The long-term benefit of sex outlined above should mean that those species that do become parthenogenetic should be more prone to extinction than those that retain sex. And indeed, this appears to fit pretty well with the distribution pattern of parthenogenesis seen in the animal kingdom. Parthenogenesis is found in a very wide array of taxonomic groups and has certainly appeared independently many times, but there are very few taxa at the higher levels (from family up) that are exclusively parthenogenetic (the bdelloid rotifers are a notable exception). Furthermore, most parthenogenetic taxa have derived from sexual taxa fairly recently. So it seems that parthenogenesis does arise fairly frequently in evolutionary time and when it does arise, the short-term advantage through the nonproduction of needless males allows it to spread. But parthenogenetic populations thereafter are more prone to elimination because they cannot compete with sexual populations in the long term.

Here then we have a viable reason for the maintenance of sex and the rarity of secondarily parthenogenetic organisms. However, this long-term, group-selectionist argument cannot be invoked to explain the origins of sex, for when sex first evolved immediate advantages must have been required. I will consider these matters again in relation to the evolution of sexual reproduction in the final chapter of this book.

CHAPTER 2

The Sex Ratio

> . . . all true power resides in the female principle. The male had served only one brief useful purpose; for the rest of his life he was a painful and costly parasite.
>
> —John Wyndham, *Consider Her Ways*

Summary

The aim of this chapter is to examine why most sexually reproducing organisms have a 1:1 sex ratio and outline theories to explain exceptions to this ratio.

Most sexually reproducing organisms produce equal numbers of males and females. The reasons that this is so are examined in this chapter. A divine explanation and a mechanistic explanation dependent on the segregation of sex chromosomes are rejected in favor of a selective explanation proposed by Darwin and Fisher. This explanation is based on a selective advantage accruing to individuals that produce the rarer sex, whichever sex this may be. Under this theory, the only stable equilibrium occurs when there is not a rarer sex, that is, when the sex ratio is 1:1.

In some circumstances selection to maintain a 1:1 sex ratio can be outweighed by other selective factors acting on individuals in particular situations. These selective factors are outlined. In some cases selective pressures on individuals in a population to produce progenic sex ratios other than 1:1 can bias population sex ratios toward one sex.

Finally, changes in the sex ratio of a cohort of individuals may occur during their lives. This gives rise to a variety of different types of sex ratios related to particular life-history moments. These are defined.

Sex Ratio Equality

Are Males Parasites of Females?

In chapter 1, we, albeit briefly, considered the history of reproduction on Earth, from asexual beginnings to sex and, in some instances, back again. In so doing, we have seen that the essential difference between males and females is gamete size and noted that this leads to differences in the way selection acts on the sexes, to a conflict of interests between the sexes, and to the evolution of many varied and extraordinary sexual strategies. Through this discussion of sex, there is a thread of feeling that females are more important than males. In many species lacking paternal care, for example, a single male, producing vast numbers of cheap sperm, has the ability to mate with and fertilize many females. From the population point of view, it could be argued that the resources available to a species in the environment would be better used if there were an extreme excess of females to males, as long as there were still enough males to fertilize all the eggs of all the females.

Some have taken this argument further, proposing the view that in species with no paternal care, males, or at least male genes, are parasitic on females. Females frequently contribute all the cytoplasmic resources to the newly fertilized zygote, males simply contributing their genes in the sperm nucleus. Of course, against this, in sexually reproducing species, almost by definition the male genes are necessary, for they restore the number of chromosomes in the zygote, thereby making a significant contribution to the female's offspring. Yet there are sufficient parthenogenetic species in most major taxa to show that males are not essential to restore or maintain ploidy levels.

The Ubiquity of the 1:1 Sex Ratio

Irrespective of whether males should be thought of as parasitic on females or not, it is a fact that in most wholly sexually reproducing species of animal, the primary sex ratio is very close to 1:1. Equal numbers of male and female zygotes are produced. This presents us with a fundamental biological question. Why, when there appear to be good reasons for increasing the ratio of females produced compared to males, do most sexual species produce approximately as many males as females?

A Divine Explanation

I will begin this section by very briefly rejecting two possible explanations. The first explanation of the equality in the sex ratio is that it was, long ago, divinely determined. According to the Bible, God initially created a man, Adam, and from him, a woman, Eve. The sex ratio created for humans was 1:1. Later, at the time of the flood, God set an equality of the sexes for all other organisms that lived when He told Noah what to bring onto the ark.

> And of every living thing of all flesh, two of every sort shalt thou bring into the ark, to keep them alive with thee: they shall be male and female.
>
> Of fowls after their kind, and of cattle after their kind, of every creeping thing of the earth after his kind, two of every sort shall come unto thee, to keep them alive. (Book of Genesis, chapter 6, verses 19–20).

Following this initial divine determination of the sex ratio, the current 1:1 sex ratio in many organisms could simply be interpreted as maintenance of the status quo.

Yet, an explanation must then be found for the exceptions to this numerical equality of males and females. The social insects, with their preponderance of females, those species that forgo sex in which males are unknown, the hermaphrodite species in which both male and female reproductive parts coexist, cannot have been gathered into the ark. The divine explanation as described in the Bible is insufficient. We must seek an explanation based on biological observation.

The Sex Chromosomes

A second explanation may be based on the sex chromosomes of most sexually reproducing organisms. As will be discussed in detail in chapter 3, sex determination in most species depends on a single pair of chromosomes. Individuals of one sex carry two copies of the same chromosome, while those of the other sex carry one copy of this chromosome and a single copy of another, usually smaller chromosome. In humans and most, but by no means all, other species, females carry two X chromosomes and males carry one X chromosome and one Y chromosome. In some groups, such as birds and Lepidoptera (butterflies and moths), this arrangement is reversed, with males carrying two similar sex chromosomes (called Zs for distinction) and females carrying one Z and a smaller W chromosome. Taking species with XX females and XY males for illustration, all the sex cells produced by females will carry an X chromosome, while half the sperm produced by males will carry an X and half will carry a Y. Therefore, at fertilization, half the eggs will be fertilized by X-bearing sperm and so will have two Xs, one from their mother and one from their father. These will be daughters. The other half of the eggs, with their X chromosome from their mother, will be fertilized by Y-bearing sperm and so will be sons. The 1:1 sex ratio can then be seen as a simple consequence of the mechanics of the most common type of sex determination, that depending on XX/XY (or ZZ/ZW) sex chromosomes.

There are two problems with this argument. First, it is feasible that rather than the sex ratio being a consequence of the mechanics of sex chromosomal sex determination, the commonness of the XX/XY or ZZ/ZW systems may simply be a consequence of the fact that these mechanisms produce the selectively optimum sex ratio.

Second, as we shall see again in chapter 3, there are many other systems of sex determination. In some species, sex is determined by single genes or by whether sex cells from the female are fertilized or not. Both the abiotic and biotic environments can act as sex determination systems, through egg incubation temperature, or chemicals produced by other members of a species in the vicinity, or even other species living within a host. Furthermore, the sex ratio is frequently distorted from parity, even in species with "normal" sex chromosomal sex determination. As Robert Ricklefs (1990) remarked:

> There are two remarkable things about the sex ratio. First, so many populations have nearly equal numbers of males and females. Second, there are so many exceptions to this rule. (p. 585)

If there were good evolutionary reasons why the sex ratio should not be 1:1, then the processes of natural selection should be able to mold the sex determination systems of species to the sex ratio that is most beneficial to the members of that species. So we must seek a selectionist explanation for the commonness of the 1:1 sex ratio.

Fisher's Theory

A selectionist explanation is available. Fundamental to this explanation is the phrase used above: "the sex ratio that is most beneficial to the members of that species." In this phrase is the realization that evolution through natural selection acts first and foremost on the individual, not the population or species. The individuals that are most fit are those that favor equality in the sex ratio. Fitness in these terms is not simply survival ability: it is also the ability to reproduce and in doing so give issue to offspring that have a high probability of survival and reproduction. Thus, in considering fitness, we must look at success across at least three generations, the success of the parental generation depending on that of itself, its progeny, and the fitness of the progeny that these produce.

Taking this view, we may consider a population of some outbreeding species in which the sex ratio is distorted toward females; say, there are two females to each male. We will consider three types of pair. Pair *A* is "normal" in that equal numbers of males and females are produced. We may assume that two of these progeny, one male and one female, will survive to reproduce. The female will mate once and produce two progeny that survive to reproduce. However, the male will mate with two females due to the biased sex ratio in the population and each of these females will produce two surviving progeny. Pair *A* will thus produce six grandchildren. Pair *B* is abnormal, in that, for some reason, it produces only daughters. Again, two offspring, both female, survive to reproduce. Each mates once and produces two surviving offspring. Pair *B* thus produces four grandchildren. Pair *C* produces only sons. Again two survive, but because of the 2:1 female:male population sex ratio, each of these males mates with two females. These four females each produces two surviving progeny, so pair *C* produces eight grandchildren. Pair *C*, the pair producing the sex that is deficient in the population is thus more successful than the normal pair, pair *A*, which, in turn, is more successful than the pair producing the common sex, pair *B*. If the population sex ratio were male rather than female biased, pair *B* would be the most successful and pair *C* the least. This argument can be seen in summary in Box 2.1.

An alternative way to view the concept was first used by Sir Ronald Fisher in 1930, when he wrote that every product of a sexual union has exactly one mother and one father. This being the case, it is easy to see that if a population of 100 males and 200 females produce 1,000 offspring, each male will, on average, contribute ten sets of genes, while each female will contribute only five sets. Again we see that there is greater value in the rarer sex.

The hypothesis that selection favors the rarer sex is usually attributed to

Box 2.1 Synopsis of the argument, attributed to R. A. Fisher (1930), that the sex ratio will tend to 1:1 in most organisms. (From Hamilton 1967.)

1. Suppose male births are less common than female births.
2. A newborn male then has better mating prospects than a newborn female, and therefore can expect to have more offspring.
3. Therefore parents genetically disposed to producing males tend to have more than average numbers of grandchildren born to them.
4. Therefore the genes for male-producing tendencies spread, and male births become commoner.
5. As the 1:1 sex ratio is approached, the advantage associated with producing males dies away.
6. The same reasoning holds if females are substituted for males throughout. Therefore 1:1 is the equilibrium ratio.

Fisher in his monumental book *The Genetical Theory of Natural Selection* (Fisher 1930). However, as Anthony Edwards (1998) has documented, the basic argument was contained within the first edition (1871) of Charles Darwin's second great book on evolution, *The Descent of Man.* The argument is clearly given and, because the first edition is not always available to readers, it is worth quoting the relevant passage:

> Let us now take the case of a species producing from the unknown causes just alluded to, an excess of one sex—we will say of males—these being superfluous and useless, or nearly useless. Could the sexes be equalised through natural selection? We may feel sure, from all characters being variable, that certain pairs would produce a somewhat less excess of males over females than other pairs. The former supposing the actual number of the offspring to remain constant, would necessarily produce more females, and would therefore be more productive. On the doctrine of chances a greater number of the offspring of the more productive pairs would survive; and these would inherit a tendency to procreate fewer males and more females. Thus a tendency towards the equalisation of the sexes would be brought about. (p. 316)

In his explanation, Darwin uses the likelihood of finding a mate rather than the number of offspring an average individual of either sex contributes to the next generation. Otherwise his argument is virtually the same as Fisher's. Furthermore, as Edwards (1998) notes, the idea is pursued by other evolutionary biologists following Darwin, such as Düsing (1884) and Cobb (1914).

It is of interest that Darwin withdrew this argument in the second edition (1874) of *The Descent*, writing

> I formerly thought that when a tendency to produce the two sexes in equal numbers was advantageous to the species, it would follow from natural selection, but I now see that the whole problem is so intricate that it is safer to leave its solution to the future. (p. 399)

This comment from the second edition has been widely quoted (e.g., Godfray and Werren 1996) and has tended to eclipse the fact that in the first edition of *The Descent*, Darwin hit the nail on the head.

This demonstration of the evolutionary stability of sex ratio equality based on the argument that the sex ratio is usually subject to negative frequency dependent selection has been hailed as the most celebrated argument in evolutionary biology (Edwards 1998). Fisher's (and Darwin's) argument was pivotal to almost every aspect of evolutionary considerations of population biology. It was the first example of an evolutionary stable strategy (ESS) and few of those described since can compete with it in elegance and simplicity. It is also a strong argument favoring the view that selection most often acts on individuals rather than groups. Of pertinence to this book, it is the theory that predicts that the normally expected sex ratio is equality and makes those species in which population or family sex ratios are distorted of interest to evolutionary biologists. Thus, it provides the baseline for a series of studies conducted over the last 35 years looking at the reasons why, in some species, the numerical parity between the sexes is flouted.

Biased Sex Ratios

Sex ratio biases, at the most general level, are of two kinds: biases in the ratio of male and female progeny produced by individuals, and the overall bias in the sex ratio in a population. The relationship between the two is not always straightforward. If for example, a proportion of the individuals in a population have a tendency to produce more daughters than sons, this may lead to a female-biased population. However, the bias produced by these individuals may be compensated for by other individuals that have a propensity to produce more sons than daughters. The 1:1 population sex ratio may therefore be the outcome of either all the reproducing individuals in the population producing equal numbers of male and female offspring, or of individuals in the population that produce a wide variety of sex ratio biases in their offspring, with the biases in favor of sons or daughters balancing each other out. Consequently, a 1:1 population sex ratio does not predict that all reproducing members of the population are producing an equality of the sexes among their progeny. Conversely, a biased population sex ratio does predict that at least some individuals in the population are producing a biased sex ratio.

It is important to recognize the distinction between population sex ratios and the sex ratios produced by individuals because biases in these two types of sex ratio may result from different types of selection.

The Trivers-Willard Hypothesis

To get a feel for the types of argument involved in explaining biases in the sex ratios, it is perhaps easiest to begin with some of the hypotheses proposed to explain abnormal sex ratios produced by individual animals similar to ourselves, that is, mammals. Among vertebrates, there are many reports of biased sex ratios in families. In the primates, for example, studies have reported that birth sex ratio is frequently influenced by the social position of mothers. In these cases there is no general pattern, high-ranking mothers of some species producing more sons than daughters (spider monkeys, Barbary macaques), other species producing the opposite bias (bonnet macaques, rhesus macaques), while in some, there is no significant bias in birth sex ratio (stump-tail macaques). Trivers and Willard (1973) argued that females in good condition would gain by producing sons rather than daughters. Their argument was that these females, by dint of their own peak condition, could provide greater investment for sons than could weak females, thereby producing high-quality sons. In doing so, these females would benefit in the long run because successful males will produce more grandchildren than daughters can. Extensive studies of red deer, *Cervus elaphus*, support the Trivers-Willard hypothesis.

Red deer stags compete in the rut for harems of females. Only large males are successful in collecting and protecting harems for a significant period of time, thereby gaining paternity. Because the rate of development of red deer calves is dependent on the level of resources provided by the mother, a female will only increase her long-term fitness by having a son if he is ultimately large enough to be successful in the rut. For a female in good condition with ample resources to deliver to her calf, producing a strong son will increase her fitness substantially if he is able to gather and hold a harem in maturity. However, if a female is in poor condition, a son is unlikely to contribute to her long-term fitness, since he is unlikely to secure any females. The differences in the fitnesses of daughters of good and poor condition females are much less because even poorly resourced females may have some reproductive success. Thus, the greater variance in reproductive success of males compared to females will lead selection to favor the production of sons in high-ranked (by condition or circumstance) females and of daughters in low-ranked females. Painstaking behavioral observations showed that high-ranked hinds carrying ample fat reserves were indeed more likely to produce sons than daughters (Clutton-Brock et al. 1984). By contrast, those with more impoverished fat reserves tended to produce daughters. The difference in reproductive success of sons born to high-ranked compared to low-ranked females was greater than the difference in the reproductive success of daughters in a similar comparison. Thus, the ability of female red deer to influence the sex of their offspring on the basis of their own rank and condition is adaptive.

This idea was widened to any situation in which the reproductive success of male and female progeny varied among parents. For many parasitoid flies and wasps, adult size depends on the size of the host that one develops within. Female parasitoids gain a greater fitness benefit from developing in a high-quality host that will provide above-average resources than do male parasitoids because high-grade resources are needed to develop ovaries and mature eggs. The result is that, theoretically, mothers will maximize their fitness by depositing female eggs into high-quality hosts and male eggs into low-quality hosts.

Many observations across a wide range of parasitoid wasp and fly species have shown that the sex ratio is correlated to host size. For example, in the pteromalid wasp, *Lariophagus distinguuendus*, which parasitizes weevils living in cereal grains, the sex ratio from small hosts (with feeding tunnel less than 1mm wide) is male biased. Larger hosts produce female-biased sex ratios, with hosts over 1.2mm producing 85–90% females (Charnov et al. 1981).

Advantaged Daughters

Given these and other similar cases, how do we explain instances in which high-ranked females produce more daughters than sons? One possible explanation has been termed the *advantaged-daughter hypothesis*. As daughters of some species remain in their natal social group and inherit the social rank of their mothers, it may be beneficial to high-ranked females to have daughters that will maintain their high social position. As males of these species migrate from their natal group before they are sexually mature, they derive little benefit from the social position of their mothers.

Local Resource Competition Hypothesis

An alternative explanation for this type of pattern is based on the reason that low-ranked females produce more sons than daughters. Named the *local resource competition hypothesis*, this theory proposes that when there is competition for food, high-ranking females will harass the daughters of low-ranked females, thereby gaining more food and reinforcing their dominance over these subordinate females. In such circumstances a low-ranked female may achieve greater success by producing sons that migrate away from the natal group.

Local Mate Competition

The explanations suggested to account for sex ratio biases related to maternal rank or condition in mammals have been shown to be relevant to many species of birds and some other vertebrates. However, these ideas are likely to be less pertinent to invertebrate species, as social interactions in the invertebrates tend to be less sophisticated than are those in vertebrates. However, biases from the 1:1 sex ratio also occur in the invertebrates. Indeed, it was in trying to explain "extraordinary sex ratios" in a variety of invertebrates, particularly insects and mites, that Bill Hamilton wrote his hugely influential paper titled "Extraordinary Sex Ratios," in 1967, a paper that kick-started recent interest in sex ratios. In this paper, he introduces the idea of local mate competition.

Hamilton knew that the sex ratio of many insect and mite populations was strongly female biased and that brother-sister matings were common in these species. These two facts led him to develop the theory of *local mate competition*. The essence of this theory is that if siblings of one sex compete with each other more than siblings of the other sex do, the fitness of the more competing sex will, for an average individual, be lower than for an individual of the other sex. Selection will then act on the parental generation to produce more of the sex that suffers less competition and thus has higher fitness.

The easiest cases to understand are perhaps those parasitoid insects that lay clutches of eggs in their hosts. Many of these insects are haplo-diploid, females being diploid as a result of normal fertilization, and males being haploid, the result of unfertilized eggs. Two unusual factors have to be understood to explain the bias in sex ratio seen in many of these species.

First, the control of the sex ratio in such organisms appears to be more directly under the influence of the female parent than in normal diploids, for it is determined simply by whether or not she allows each of her gametes to be fertilized before it is laid.

Second, these haplo-diploids are unusual in another way. Inbreeding depression, or the loss of fitness due to the phenotypic expression of harmful recessive genes, is generally low. Why? Because of the haploidy of the males. As males carry only one copy of each gene, any deleterious recessive mutations would be expressed whenever they occurred in a male, for they could not be "hidden" by a dominant allele on a homologous chromosome, as they could be in the diploid females. Thus, deleterious recessive genes, unless only active in females, would be expressed in males, selected against, and so expunged from the population as soon as they arose.

Given the low level of inbreeding depression in haplo-diploids, mating with relatives is no longer a problem. Here, then, when females lay a batch of eggs in a host, if males have the ability to mate many times, there is little reason to produce more than one male. As long as a male hatches before any of his female siblings hatch from their pupae and disperse, he will be in a position to mate sequentially with each of his sisters in the batch as she ecloses. And, indeed, this is the pattern seen in many haplo-diploid insects. Hamilton (1967) lists 25 such cases, including parasitoid wasps and mites, fig wasps, and the moth ear mite. Many other similar instances have subsequently been described.

Professor John Werren (1980), studying the jewel wasp, *Nasonia vitripennis*, elegantly tested Hamilton's prediction. This wasp parasitizes the puparia of a wide variety of flies. Females lay variable numbers of eggs into fly puparia, the number being affected by both the size of the host and whether the puparium has previously been parasitized. In fact, superparasitism, whereby two or even more females oviposit into the same host, is fairly common in this species. Ovipositing females have the ability to detect whether a host has been attacked previously. In instances of superparasitism, the progeny from both females develop together and hatch at the same time. Typically, males hatch before females. All exit from the puparium through a single hole and the males then compete for a position close to the hole through which the females will hatch. The males then mate with the females as soon as they issue from the exit hole. Females only mate once and then disperse to find hosts in which to lay their eggs.

Werren saw that the optimum sex ratio produced by the first and second female to lay in a host would be different. The first female to lay should produce just enough males to ensure that all the females were inseminated. The average he found was about one male for every nine or ten females produced. The second female should produce a greater proportion of males because these will have the opportunity to mate with the many daughters of the first female. Ideally, the precise sex ratio produced by the second female depends not only on whether the host has previously been parasitized, but also on the relative numbers of eggs laid by the two females. If the first female has laid many eggs into a host, Werren predicted that the second female should lay only a few eggs and that all should be male. Conversely, if the first female has laid few eggs, so that there is scope for the second female to lay a larger number of eggs into the host, the proportion of males should be lower; otherwise, the second female's sons will wastefully compete for matings among themselves. Thus, as the number of eggs laid by the second female increases relative to the number laid by the first female, the proportion of males in the progeny of the second female should decline. When this system was tested experimentally, Werren's predictions were borne out by his observations with great precision.

The case of *N. vitripennis* not only lends strong support to Hamilton's local mate competition theory, it also bears out Fisher's theory. Fisher's view that parents should invest equally in male and female offspring assumes an outbreeding population. Werren's work showed that the sex ratio could deviate from parity, in a predictable direction and to a predictable extent, if this assumption of outbreeding were violated.

Hamilton's 1967 paper has spawned much research into biased sex ratios. Many studies have supported the local mate competition hypothesis. However, much of the empirical evidence generated by these studies is complex and its interpretation is not always straightforward. To give an example of two data sets that initially appear contradictory, we may consider two comparative studies of the sex ratios in Hymenoptera. King and Skinner (1991) looked at the sex ratios produced by two species of *Nasonia*, *N. vitripennis* and *N. giraulti*, the former of which has wingless males and the latter winged males. They found that the winged male species, *N. giraulti*, produced more female-biased sex ratios than *N. vitripennis*. In the second study, West and Herre (1998) looked at the sex ratios of 17 species of fig wasp, some of which have winged males that disperse with females, while others have apterous males that die in their natal fruit. Here species with winged males produced less female-biased sex ratios, the opposite of the pattern seen in *Nasonia*. The resolution of this apparent contradiction lies in the timing of mating. In the fig wasps with winged males, many females that disperse from their natal fruit can mate after dispersing. Males that also disperse at this time thereby have some future mating opportunities. Their production may thus increase the long-term reproductive success of their parents. Conversely, in *N. giraulti*, a very high percentage of females are inseminated before they leave their host fly puparia. Males that disperse are thus unlikely to find virgin females (Drapeau 1999).

It is now known that many species of parasitoid, in which siblings develop in the same host, produce highly female-biased sex ratios and that females of some of these produce clutches of eggs that are exclusively female, with a single exception of one unfertilized male egg.

This pattern is taken to extremes in the mite *Acarophenax*, in which mothers produce live young. Before the female releases a clutch of young, usually about 15 in number, the single male in the clutch of eggs will have hatched and mated with each of his sisters while still inside his mother. He then dies before being born, so that the female just gives birth to living daughters that have already been fertilized.

In these cases of sex ratios biased toward females due to local mate competition, we hear echoes of Rincewind's Nirvana. However, if the idea of being one male among many times that number of females is attractive to some readers, presumably the men, it is perhaps worth pointing out that such a situation may have its drawbacks. Take the case of a small black wasp called *Dinocampus coccinellae* which parasitizes adult ladybird beetles (Plate 1b). Imagine a male of this species, which, having developed inside a ladybird, has burrowed out, made his cocoon between the immobilized legs of the still living host (Plate 1c) and emerged into adulthood. This male is one of a very few, only a dozen males of this species having been reported and the sex ratio is probably well in excess of 1,000 females for every male (Geoghegan et al. 1998). Our male should be in clover, but, in fact, evolution has left him behind, for female *D. coccinellae* reproduce parthenogenetically. Although males on encountering a female will engage in courtship behavior typical of related sexually reproducing wasp species and will go as far as attempting to mount the female, she will rebuff all mating advances by the male who is no longer necessary to her.

The Influence of Relatedness on the Sex Ratio

The cases of biases in the sex ratio outlined above all concern biases in the sex ratios produced by individuals. Furthermore, the selective rationales underlying these biases all depend to some extent on interactions between relatives, whether these be mating between siblings, competition between relatives or the resourcing of offspring by mothers. Such biases, while essentially the result of selection acting at the individual level, frequently impact on the population sex ratio. Thus, for example, Torres and Drummond (1999) have reported that populations of Blue-footed Boobies produce male-biased cohorts during poor breeding seasons. They argue that this is adaptive, since daughters are more costly to produce than sons. By contrast, wild horses in the Misaki region show a female-biased population sex ratio (Khalil and Murakami 1999). Among the invertebrates, a great many haplo-diploid parasitoid wasps exposed to local mate competition have strongly female-biased population sex ratios because most, or in some cases, all females produce fewer sons than daughters.

It is apparent that selection acting on individuals in specific circumstances may cause these individuals to produce more of one sex than the other. Cases when these individual sex ratio biases produce population sex ratio biases, and appear to have done so over long periods of time, show that there are circumstances when the evolutionary stability of the 1:1 sex ratio can be permanently disturbed.

Primary, Secondary, and Tertiary Sex Ratios

One point that perhaps needs some clarification before proceeding, is what is actually meant by the term *sex ratio*. In its simplest form, of course, we simply mean the ratio of males to females. This is often given as simply the proportion of a group of organisms under consideration that is male. This is the practice I will follow unless I specify otherwise. Thus, in populations that are female biased, the sex ratio will be given as a figure less than a half, or SR < 0.5. However, we need to go a little further, as the sex ratio of a group of organisms may change with the age of that group. Sex ratio may therefore be split into three categories: primary, secondary, and tertiary, on the basis of critical times in the life cycle. Formally, the primary sex ratio is defined as the proportion of males at zygote formation; that is, directly after fertilization. The secondary sex ratio is the proportion of males at birth, and the tertiary sex ratio is that at sexual maturity. Unfortunately, many writers have ignored these formal definitions and have been confused in their usage. Thus, it has become common practice to refer to sex ratios assessed at birth as primary sex ratios and all later sex ratios as secondary, deleting the tertiary sex ratio completely. Others have defined the primary sex ratio as that at either conception or birth (Bourke and Franks 1995), which are not the same thing at all, while Trivers (1985) extends the primary sex ratio in species with parental care to the end of the period of parental investment.

In this book, when discussing mechanisms that influence the sex ratio, I shall refer to those that act before or at zygote formation as changing the primary sex ratio. Those that change the sex ratio after zygote formation but before reproductive maturity is attained will be referred to as altering the secondary sex ratio, irrespective of whether they act before or after birth. Distortions of the sex ratio that act once reproductive maturity has been attained will be considered tertiary.

The Problem of Birth

I hope that this usage is clear, for it allows me to sidestep another problem of definition that occurred to me while considering these various sex ratio categories. That is simply that I do not know what birth is. The moment of birth of a human is easy to pinpoint. However, in egg-laying organisms, does birth occur when the mother lays the egg, or when the egg hatches? Worrying over this problem, I sought help from my peers in the Genetics Department in Cambridge. Of 84 people who responded to a simple e-mail inquiry, the split was 21 in favor of birth occurring as the egg is laid, with 63 voting for the moment of hatching. Despite this majority for hatching, the split decision highlights that there is a difficulty here, and if birth occurs on hatching, then common usage of the term "live birth" for viviparous species is a misuse. The problem becomes even more problematic if extreme examples, such as *Acarophenax*, are considered. In describing the mating in this mite, therefore, I have already made an apparently nonsensical statement resulting from this semantic difficulty, writing (p. 46) that the males of this species die before they are born. Indeed, in these remarkable mites both males and females actually manage to mate before they are born.

Although this is something of a semantic difficulty, as we shall see, the evolutionary influences on the sex ratio that act at different stages in a species' life history may be very different. The ease with which the sex ratios of a species can be manipulated by parents, offspring, or, indeed, other organisms, depends to some extent on the sex determination system of that species. It is to the many and varied ways in which the sexes of organisms are determined that I will now turn.

CHAPTER 3

Sex Determination

For spirits when they please
Can either sex assume, or both; so soft
And uncompounded is their essence puree . . .
in what shape they choose,
Dilated or condensed, bright or obscure,
Can execute their aery purpose.
—John Milton, *Paradise Lost*

Summary

Most sexually reproducing organisms produce two types of gamete that differ in size. By definition, the primary sex organs, or gonads, that produce the larger gamete are female, while those that produce the smaller gamete are male. In most, but not all organisms, male and female gonads are on different individuals, males and females, respectively. The aim of this chapter is to describe and examine the many and varied ways that the sex of an individual is determined.

The chapter first considers a variety of situations in which individuals may be partly male and partly female. Thereafter, a variety of sex determination mechanisms are considered.

In many organisms, sex chromosomes determine sex. In some, specific genes on the sex chromosome only possessed by one sex are of paramount influence. In others, the number of copies of the sex chromosome that are present in both sexes, in relation to the number of copies of autosomes, is the critical sex determinant. Less common types of chromosomal sex determination and single-gene sex determination are then discussed. Finally, a variety of environmental factors that influence sex determination in particular species are described.

Throughout the chapter, the evolutionary causes and consequences of the various types of sex determination mechanism are visited.

Sexual Mosaics, Intersexes, and Hermaphrodites

Most sexually reproducing, multicellular organisms have two types of sex organs. In some species the male and female sex organs are on the same individual (monoecious); in others, such as us, they are normally on different individuals (dioecious). The term *hermaphrodite* is used for individuals that bear both male and female reproductive organs. Thus, monoecious species, such as a great many plants and many types of invertebrate are hermaphrodite. Indeed, of 28 animal phyla, 20 have at least some hermaphroditic species, and seven (sponges, the entoprocts, the bryozoans, the flatworms, the arrowheads, the gastrotrichs, and the comb jellies) are exclusively hermaphrodite. Three other phyla are largely hermaphrodites (anenomes and corals, sea slugs and pulmonate snails, and gnathostomulids, leeches, and earthworms). In addition, some species have both hermaphrodite and single-sex individuals occurring, an example being the nematode worm *Caenorhabditis elegans*, which is usually hermaphrodite, but in which males also occur when one of the normal two sex chromosomes is lost.

In some normally dioecious species, individuals with both male and female primary sex organs occur occasionally. Such individuals are most often recorded in species with obvious secondary sexual differences, such as color or morphological differences, and are often first noted because an individual has both some male and some female attributes. Individuals that appear to be made up of some male and some female tissues are called *sexual mosaics*. Such individuals can be differentiated from intersexes that exhibit both male and female traits and some traits that are intermediate between the two, yet do not have genetically different parts.

The most spectacular sexual mosaics are those in which the split between the two sexes is down the middle, one side being male, the other female. Many such examples have been recorded—most in insects. This is partly because sex determination in many insects provides a pathway in which plausible errors may lead to bilateral mosaicism, and partly because many insects show secondary sexual dimorphism, and have long been collected and studied by entomologists, so that when they occur they are detected.

The term *hermaphrodite* is commonly used for abnormal dioecious organisms that are part male and part female. This common usage should be distinguished from the scientific definition of a hermaphrodite, which stipulates that the individual in question must bear both male and female reproductive organs. Individuals that have some male tissues and some female tissues, but do not carry both types of sex organ should be referred to as sexual mosaics, not hermaphrodites.

The word *hermaphrodite* derives from Hermaphroditus, the child born to the goddess Aphrodite and the man Hermes. Hermaphroditus became captivated by the nymph Salamacis and joined with her to become one part-male, part-female, being. Some Greek sculptures of human hermaphrodites show them split down the middle, one side male and the other female. The Hindu deity Ardhanarisvara, who is half-god, half-woman, is shown as female on the left side (most obviously because of a single breast) and male on the right (with half a moustache). Even more intriguingly, a seventeenth-century report from Italy describes an apparent case of bilateral sexual mosaicism, with a man giving birth to a child.

> A weird happening has occurred in the case of a lansquenet (soldier) named Daniel Burghammer. . . . When the same was on the point of going to bed one night he complained to his wife, to whom he had been married by the Church seven years ago, that he had great pains in his belly and felt something stirring therein. An hour thereafter he gave birth to a child, a girl . . . He then confessed on the spot that he was half man and half woman. . . . He also stated that . . . he only slept once with a Spaniard, and he became pregnant therefrom. This, however, he kept a secret unto himself and also from his wife, with whom he had for seven years lived in wedlock, but he had never been able to get her with child. . . . The aforesaid soldier is able to suckle the child with his right breast only and not at all on the left side, where he is a man. He has also the natural organs of a man for passing water. . . . All this has been set down and described by notaries. It is considered in Italy to be a great miracle and is to be recorded in the chronicles. The couple, however, are to be divorced by the clergy.
> —From Piadena in Italy, the 26th day of May 1601 (von Klarwill, 1924)

This case is surprising and must be treated with considerable skepticism, since current understanding of sexual development in mammals predicts that strictly bilateral sexual mosaics do not, indeed cannot, occur in humans. This is because of the influence of hormones on the development of sexual characteristics in mammals. Because these hormones are carried around the whole body in the blood, it is difficult to envision a mechanism for the development of a milk-producing breast just on one side. As mentioned above, bilateral sexual mosaics do occur in some organisms, but the sex determination systems of species in which they have been recorded contrast strikingly with those in higher vertebrates such as mammals and birds. So how is the difference between males and females determined? What are the essential differences in the genetic mechanisms that lead to different pathways of development?

Humans have long been fascinated by the factors that determine sex. Many views have been put forward over the millennia. Aristotle, for example, felt that sex could be influenced at the time of conception. He advised that the probability of a couple producing a son would be increased if the marital bed were placed on a north-south axis. Anaxaogora regarded the position during copulation as being influential, sons being conceived if a couple lay on their sides during the act of sex. Later, there was a belief that if sperm derived from the right testicle, a son would result, while sperm from the left testicle would produce a daughter. This led some French aristocrats to have their left testicle surgically removed to increase the probability of having sons. The contemporary view of sex determination is based on a complex interplay between genes, chromosomes, hormones, and the environment.

Sex Chromosomes

In most, but not all multicellular, sexually reproducing organisms, the chromosome complements of males and females differ with respect to just one pair of chromosomes. The chromosomes of this pair are called the *sex chromosomes.* In one sex, both chromosomes of this pair will be similar in size (called the homogametic sex, as all gametes produced are similar with respect to the sex chromosome that they carry). The other sex will have just one sex chromosome of this type, with the second sex chromosome being obviously different—usually much smaller (called the heterogametic sex, as two different types of gametes will be produced, either with the larger or with the smaller sex chromosome). When it is the female that has two similar sex chromosomes, the sex chromosomes are called X and Y, females being XX and males XY. This is the pattern in most taxa. However other patterns are also observed. In some species, females are typically XX while males are XO, lacking the Y chromosome (see p. 62). In a few taxonomic groups, the sex chromosomes appear reversed, with females being heterogametic and males homogametic (see p. 57). In these species, the chromosomes of the homogametic males are denoted as ZZ, while those of the heterogametic females are ZW. Finally some species have sex determination systems that do not rely on sex chromosomes at all. To understand how sex is determined and consider how sex ratios sometimes become distorted, all these patterns of sex determination should be considered. It is perhaps sensible to begin with the most commonly occurring pattern, that involving XX females and XY males.

XX/XY Sex Chromosome Systems

I will start with what appears to be a very simple question: is it the presence of a Y chromosome that makes a male a male, or is it the absence of one of the X chromosomes? The answer is less simple, for it is different for different taxonomic groups. This can most easily be appreciated by considering sex chromosomal abnormalities in two well-studied organisms, humans and fruit flies (*Drosophila*).

Human Sex Determination

In humans, several abnormalities involving the sex chromosomes are known (Table 3.1). Individuals with a single sex chromosome, an X (Turner's syndrome), are essentially female. On its own, this suggests that the Y chromosome is vital to male development in humans. Corroborative evidence comes from the fact that presence of two Xs and a Y chromosome (Klinefelter's syndrome) produces a male, showing that the presence of two X chromosomes does not always produce a female. Indeed, the number of X chromosomes does not appear to be important in basic sex determination in humans for individuals with one (Turner's), two (normal), or three (called Triple X) X chromosomes are all essentially female as long as no Y chromosome is present.

More detailed analyses of embryological developmental systems have shown that the central event in mammalian sex determination is the formation of the testis. The testes and ovaries both derive from the same embryological struc-

TABLE 3.1
Sex chromosome abnormalities in humans.

Name of Abnormality	*Sex Chromosome Complement*	*Essential Phenotype*
Turner's syndrome	XO	Female
Klinefelter's syndrome	XXY	Male
X double Y	XYY	Male
Triple X	XXX	Female

ture: the gonadal ridge. In the presence of the Y chromosome, the gonadal ridge differentiates to form a testis: in the absence of the Y chromosome, an ovary forms. In males, hormones produced by the testis (e.g., testosterone and anti-Mullerian hormone) drive subsequent sexual differentiation. When these hormones are not produced, female differentiation occurs. It has been shown that a single gene, *SRY*, present on the Y chromosome, is responsible for inducing male testis formation in mammals. This gene is believed to act as a transcription factor controlling the expression of other genes, not located on the Y chromosome, in a genetic-control pathway.

The *SRY* locus is on a part of the Y chromosome that does not pair with any part of the X chromosome but it is located very close to the section that does pair. Very rarely, the *SRY* gene becomes transferred to the end of an X chromosome. The result is an individual whom cytological examination reveals to have two X chromosomes and no Y, yet is male. Existence of such individuals is of interest to sporting organizations, for they confound the normal "Barr body" sex test for athletes. This test depends on the fact that one of the X chromosomes of a female is inactivated. The inactivated chromosome does not behave like other chromosomes. Its DNA is modified by the addition of many methyl groups and it condenses into a tight structure that can be stained in microscopic examination. This structure is called a *Barr body* after Murray Barr, the Canadian geneticist who first discovered it. In a normal female, this structure can be seen attached to the inner surface of the nuclear membrane where it replicates asynchronously to the other chromosomes in all tissues except germ line tissues in which it is reactivated. Presence of the Barr body indicates the presence of two X chromosomes, one of which has been inactivated; hence its use as a sex test. However, XX individuals that have the *SRY* gene translocated onto one of their X chromosomes and are thus male, show up as female in the Barr body test.

The role of testosterone in human sex determination is evident from examination of people suffering from a syndrome known as *testicular feminization*. Once testes begin to secrete testosterone, this hormone binds to receptors on many types of cells. The testosterone-receptor complex that is formed instructs the cell to develop as a male or a female. Testicular feminization results from a mutation in a gene on the X chromosome, *tfm*, which prevents the production of the testosterone receptors. Although testes are formed so that testosterone is produced, absence of the receptors on the target cells causes embryos to switch sexes during development and produce female characteristics. However, ovaries are not produced, with the result that individuals with testicular feminization are sterile.

The importance of hormones to sex determination and development in humans is also evident from the effects that inappropriate hormone exposure can have on early-developing fetuses. If a mother is exposed to high levels of estrogen in early pregnancy, a genetically male (XY) fetus may be born with essentially female genitalia. Similarly, testosterone or progestrogen exposure in early pregnancy can lead to the partial masculinization of a genetically female (XX) fetus, the baby being born with abnormal genitalia, including a small penis.

One other point of interest related to the fact that sexual development in mammals and many other vertebrates is determined by hormonal systems is the current increase in observations of sexual-development abnormalities. These include the production of intersexes and partial hermaphrodites, reduced sperm counts, increases in sperm abnormalities, and sex organ–related cancers. These abnormalities have been attributed by some to increased exposure to so-called estrogenic chemical compounds. These comprise a diverse array of compounds, including some insecticides and emulsifiers among many others that have an active site that mimics the action of the female hormone estrogen. Levels of exposure to estrogenic compounds are known to have phenotypic effects on the secondary sexual traits in mammals. For example, in humans, sex change procedures from male to female involve administration of estrogen and other female hormones, leading to increased development of mammary glands and reduction in facial and body hair. These observations in humans, together with sexual abnormalities recorded in other vertebrate taxa, including trout in Britain and crocodilians in Florida, suggest that disruption of sex hormonal balances can have severe effects on reproductive capabilities. They also attest to the crucial role that hormonal systems play in sex determination and sexual reproduction in these groups.

Dosage Compensation

I have already mentioned in passing that in many female mammals, although not in marsupials, one X chromosome is inactivated in somatic cells, this chromosome producing the Barr body. This inactivation is one answer to a problem faced by all organisms that have two of one sex chromosome in one sex and only one of that chromosome in the other sex. The problem is this. Most genes come in pairs. Two copies are present in all individuals. More or less than two copies of genes often leads to abnormalities. Consequently, the difference in the number of copies of X-linked genes in females and males might be expected to produce difficulties unless the dose of these genes in females and males is equalized in some way. Three mechanisms seem possible (Figure 3.1). First, one copy of each X-linked gene may be inactivated so that both females and males have just one active copy. This is in fact what happens in humans and

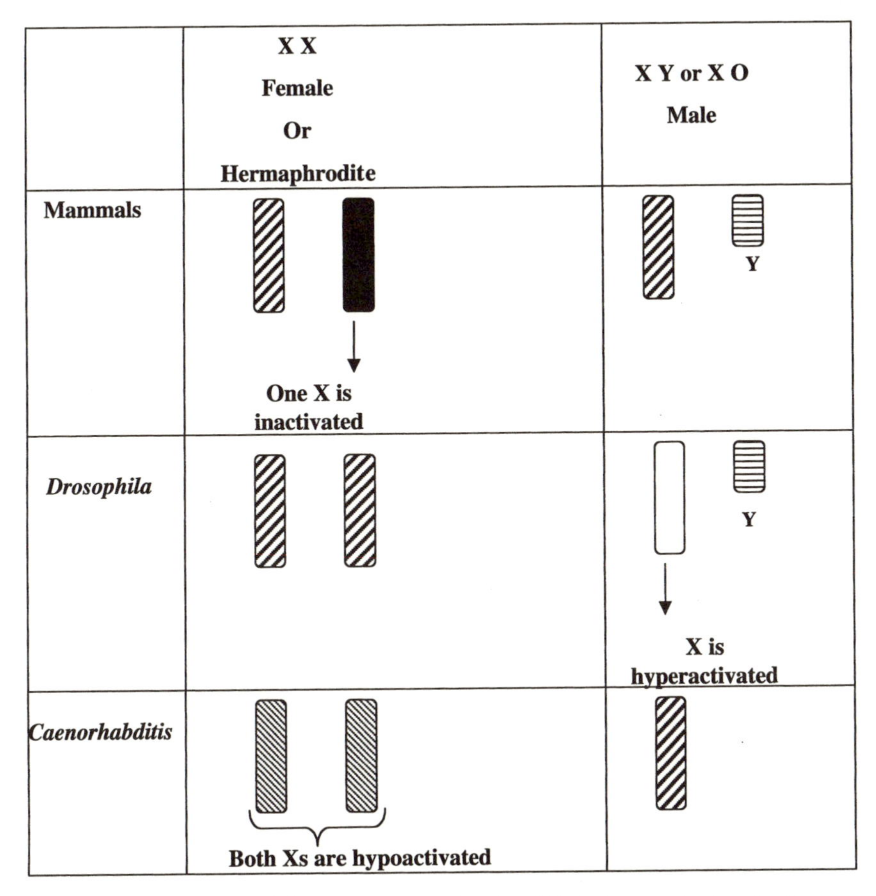

Figure 3.1 Mechanisms of dosage compensation for X-linked genes by inactivation (mammals), increased activation (fruit flies) and reduced activation (nematodes). (Derived from Snustad et al. 1997.)

most other mammals and is the reason for the inactivation of one X chromosome in females. The different dosages of the genes (two in females and one in males) have been compensated for by the inactivation of one copy in females. Second, the activation of genes on both X chromosomes may be partially repressed, as occurs in nematode worms (p. 000).

Sex Determination in Drosophila

The third mechanism is that each gene on the nonpairing part of the X chromosome in males has to produce twice the dose of its product to compensate for the two copies in females. This is the mechanism that operates in fruit flies, such as *Drosophila melanogaster*.

The mechanism of sex determination in *D. melanogaster* was first elucidated

TABLE 3.2
Chromosomal complements and sexual phenotypes in *Drosophila melanogaster*.

Number of X Chromosomes and Number of Sets of Autosomes (A)		*X:A Ratio*	*Phenotype Produced*
1X	3A	0.33	Super male
1X	2A	0.5	Normal male
2X	4A	0.5	Tetraploid male
2X	3A	0.67	Intersex
3X	4A	0.75	Intersex
2X	2A	1.0	Normal female
3X	3A	1.0	Triploid female
4X	4A	1.0	Tetraploid female
4X	3A	1.33	Super female
3X	2A	1.5	Super female

by examination of individuals with abnormal chromosome complements. Many chromosomal abnormalities have been recorded in these flies. Some involve abnormal numbers of sex chromosomes, while others involve increases in the number of sets of autosomes (nonsex chromosomes). Table 3.2 gives details of a number of known chromosomal complements in *Drosophila* together with the sex of the fly that usually results. Crucial to our question of whether it is the presence of the Y, or the lack of a second X that determines maleness, is the cytotype with a single X, no Y, and two sets of autosomes. Such individuals are essentially male. The presence of a Y chromosome is thus not necessary for male development.

Molecular genetic analysis has revealed that a gene called sex lethal (*Sxl*) plays a key role in the sex determination system of *Drosophila*. A complex set of other X-linked genes, working with factors in the cytoplasm of the fly egg, determines the level of expression of *Sxl* in a new fly zygote. If the ratio of X chromosomes to sets of autosomes (denoted as A) is equal to or greater than 1, the embryo develops as a female. If this ratio is equal to or less than half, a male is produced. Ratios intermediate between 0.5 and 1 result in intersexes, with some male, some female, and some intermediate characteristics. Here then is a genetic system that appears to involve both counting chromosomes, then working out the ratio of Xs to As, and, on the basis of the result, switching the *Sxl* gene on or off as appropriate.

The system used to calculate the X:A ratio is based on proteins that are synthesized by the mother and deposited in the cytoplasm of her eggs. These proteins interact with other proteins that are made by the embryo due to expression of several X-linked genes. When the mother's proteins and the embryo's proteins are both free to interact, the *Sxl* gene is switched on. However, the proteins from the X-linked genes are not always free. Genes on the autosomes of the embryo produce other proteins that will bind to the proteins from

X-linked genes, preventing them from interacting with the maternally contributed genes. The important question is then: Are there enough autosomal proteins to "tie up" all of the X-linked proteins (Figure 3.2)? The amount of autosomal protein will be the same in normal XX and XY flies. The sex chromosomes will have no influence here. However, because the amount of protein produced by the X-linked genes is directly proportional to the number of copies of each gene and so to the number of X chromosomes, XX flies produce double the amount that XY flies do. In XX flies, then, the X-linked genes are present in excess, and interaction with the maternal protein in their egg cytoplasm allows *Sxl* to be activated. The fly then develops as a female. In XY flies, the autosomal proteins are in excess, so there is no interaction with the maternal proteins and *Sxl* is not activated at the critical time. A male fly then results (see Gorman and Baker 1994).

The *Sxl* gene is thus the master gene that regulates sex determination in *Drosophila*. In XX individuals (or others with an X:A ratio of 1 or more), *Sxl* produces a functional polypeptide, called SXL, which regulates its own continued production and the full development of female traits. In XY individuals (or others with an X:A ratio of 0.5 or less), *Sxl* does not produce SXL (see Cline 1993 for review).

Another important feature of *Drosophila* sex determination was revealed by study of sexual mosaics; that is, individuals with some female and some male tissues. The first flies of this type that were found and studied were found to have some cells with two X chromosomes, and some with one X but no other sex chromosome (XO). These mosaics occur naturally but very rarely when one X chromosome is lost from some cell lineages during mitosis. In the mosaic gynandromorphs that result, tissues composed of XX cells develop into female structures, while those composed of XO cells produce male structures. The exact correlation between the sexual form of a tissue and its chromosomal complement shows that the sex of a cell is entirely a product of processes that operate within that cell. There is no influence from other cells. Thus, hormones that move between cells are not important here.

ZZ/ZW Sex Determination

In fruit flies, XO individuals or tissues are an exception. However, in some species, Y chromosomes are always absent, females being XX and males XO. Before discussing such species, some consideration of the differences between XX/XY and ZZ/ZW systems in two insect groups, the Diptera and the Lepidoptera, may be appropriate. Species with ZZ/ZW sex determination are often mentioned in scientific textbooks almost as an afterthought: a simple note points out that in some groups—birds, most butterflies and moths, and a few reptiles and fish—the sexes appear switched over, females rather than males being the heterogametic sex (Figure 3.3). This is excusable because the vast majority of work on sex chromosomal sex determination has involved XX/XY groups such as the drosophilids. However, some research has been conducted on sex deter-

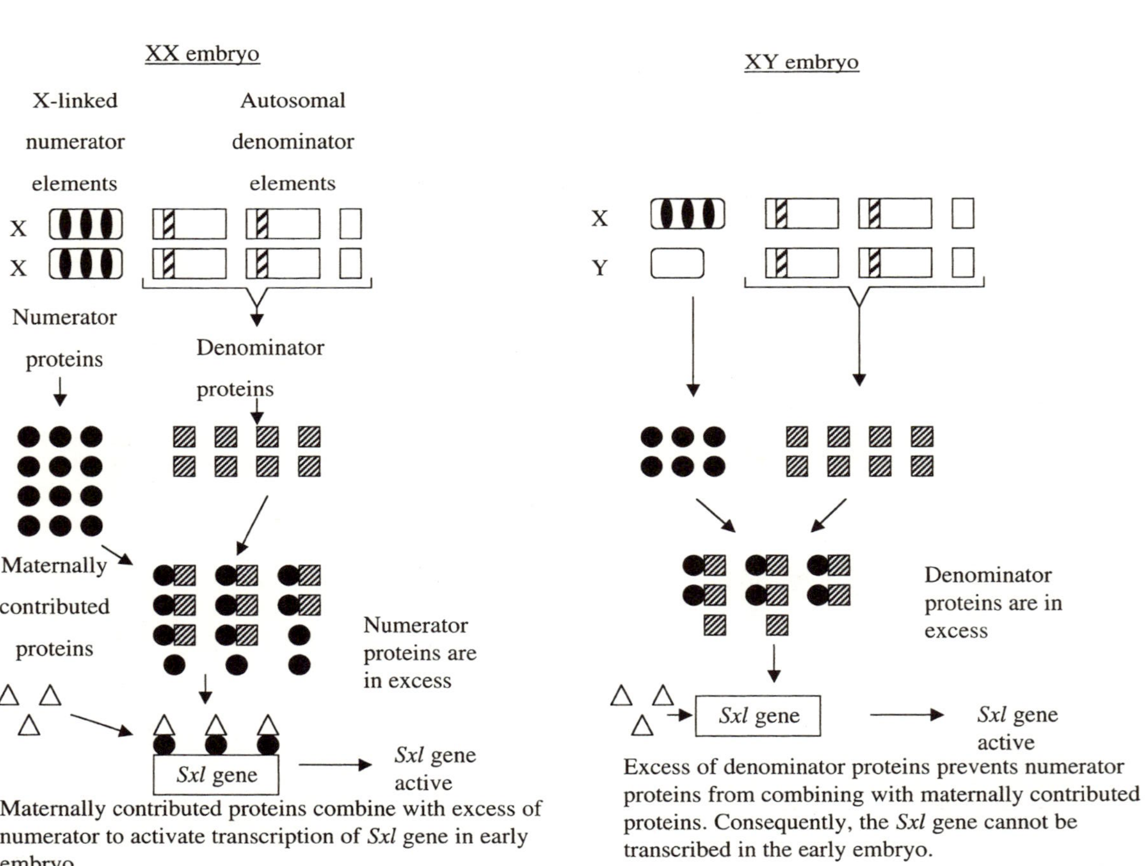

Figure 3.2 How *Drosophila* estimates the X:A chromosomal ratio. The ratio of X chromosomes is derived as a result of interactions between proteins produced by X-linked and autosomal genes. (Redrawn from Snustad et al. 1997.)

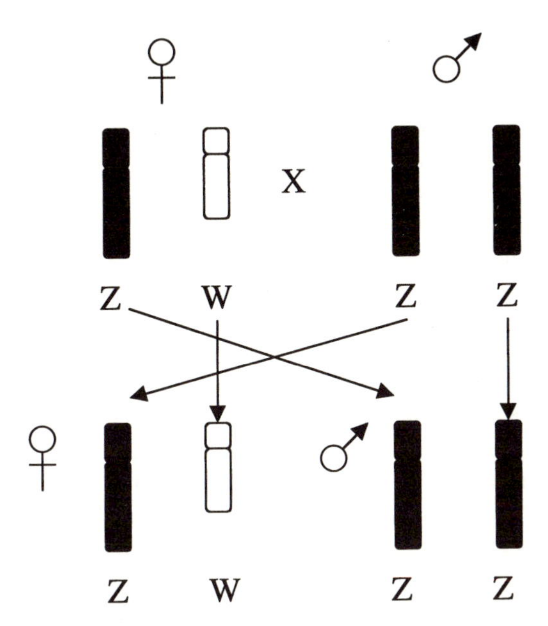

Figure 3.3 Sex determination in groups, such as birds and the Lepidoptera, in which females are homogametic and males heterogametic. The sex of offspring is determined by whether eggs from the female parent possess a W chromosome or a Z chromosome.

mination in ZZ/ZW systems, particularly in the Lepidoptera. This work is worthy of attention because it shows both interesting similarities and interesting differences in the genetics of sex determination in the drosophilids and the Lepidoptera. In addition, the heterogametry of females rather than males in ZZ/ZW species may have considerable effects on the evolutionary genetics of groups such as the Lepidoptera and birds.

Much of the work on sex determination in the Lepidoptera has concerned a species that has long been extinct in the wild, the commercial silkworm, *Bombyx mori*. As much of this discussion will involve comparison with the human and drosophilid sex determination systems mentioned above, it is worth putting *B. mori* in a genetic context against these. The haploid genome of *B. mori* contains roughly 530 million nucleotide base pairs. This means that it is over 3½ times the size of the genome of *Drosophila melanogaster* and about one-sixth the size of the human genome.

Female *B. mori* are ZW; males are ZZ. However, in other species of Lepidoptera, females known to be ZO, ZZW, or even ZZWW occur. The presence of both ZO and ZZW females suggests that the W chromosome is essential in female determination in some species (ZZW) species, but not in others (ZO species). Tazima (1964), considering a model proposed by Hasimoto in the

1930s, conducted experiments on *B. mori* strains with extra chromosomes and translocated fragments of the W chromosome. His analysis showed that the W chromosome does carry female-determining genes in this species, as proposed by Hasimoto. The positions of these genes were localized to one end of the W chromosome. In contrast to the situation in fruit flies, alteration of the ratio of sex chromosomes to autosome sets (in *B. mori*, the Z:A ratio) had no effect on the sex produced. The absence of dosage compensation has recently been confirmed by molecular genetic analysis of a Z-linked gene in *Bombyx*.

Bombyx mori is certainly not typical of all Lepidoptera. Obviously, the W chromosome in ZO species cannot determine femaleness. In addition, in chromosomally abnormal strains of *B. mori*, sexual mosaics rather than blended intersexes were produced. This is in contrast to many other species of Lepidoptera in which intersexes have been reported relatively frequently.

Another species of moth in which sex determination has received considerable attention, due largely to the painstaking work of J. Seiler over a period of fifty years, is *Solenobia triquetrella*. This is an extraordinary species for three reasons. First, some of the moths are diploid with 62 (occasionally 61) chromosomes while others are tetraploid, that is, they have four sets of chromosomes (chromosome number = 120–124). Second, some diploids reproduce sexually in the normal way, while other diploids and all tetraploids in the wild reproduce by thelytokous parthenogenesis in which females just produce daughters. Third, the sex chromosome complement of diploid females in the wild is variable, some being ZW and others ZO.

Looking at the first two features, there appear to be three types of moth: sexually reproducing diploids, parthenogenetic diploids, and parthenogenetic tetraploids. The situation is further complicated because some females have the ability to reproduce either sexually or parthenogenetically, depending on whether sperm is available for fertilization. It is perhaps worth noting here that the presence of some ZO females may have facilitated the transition to parthenogenesis. Indeed, neither diploid nor tetraploid parthenogenetic females have W chromosomes.

With respect to sex determination, it is the product of matings between sexual diploid males and parthenogenetic females that is perhaps most interesting. Given that some females are ZW and some are ZO, this is obviously one of the species in which the W chromosome does not play a crucial role in determining femaleness. Indeed, observations have shown no significant difference between ZW and ZO females in terms of viability or reproductive efficiency, suggesting that the W chromosome may have negligible effect (Seiler 1959, 1960). This speculation is endorsed by analysis of individuals with one or two extra W chromosomes. ZWW moths are female but ZZW and ZZWW moths are male. These chromosomally abnormal moths appear to be no different phenotypically from female ZW or male ZZ moths. Here, then, it appears to be the number of Z chromosomes, rather than the presence or absence of the W chromosome, that is crucial in sex determination. In captivity, sexual diploids can be induced to mate with parthenogenetic tetraploids without much difficulty. The progeny

TABLE 3.3
Chromosomal complements and phenotypes found in *Solenobia triquetrella*.

Sex Chromosome Complement	*Phenotype*
ZO : 2A	normal female or parthenogenetic female
ZW : 2A	normal female or parthenogenetic female
ZZ : 2A	normal male
ZWW : 2A	normal female
ZZW : 2A	normal male
ZZWW : 2A	normal male
ZZ : 3A	variable sexual mosaics and intersexes
ZZ : 4A	parthenogenetic female

of such crosses are of two types, normal tetraploid parthenogenetic females and triploid intersexes. These results are explained if only some eggs are fertilized and if sex determination in this species depends on the Z:A ratio, somewhat like the situation in drosophilids. If in normal sexual diploids, males have two Z chromosomes and two sets of autosomes (2Z2A) and females are Z2A, we can consider what happens when diploids are mated with tetraploids. Unfertilized eggs will develop in the same way as the eggs of unmated tetraploids. Their chromosome complement will be 2Z4A. The ratio of Zs to As will be the same as in normal diploid females. However, a meiotic product of one of these tetraploid females would be Z2A. Were this to be fertilized by a normal haploid sperm (ZA), the zygote produced would have two Z chromosomes but three sets of autosomes. The ratio between Zs and As would thus be intermediate between that for normal males and normal females, and intersexes would result (Table 3.3).

Although, on current evidence, it cannot be said of all Lepidoptera, it appears that in many Lepidoptera there is no dosage compensation related to the Z chromosomes. Studies of a Z-linked gene (the 6–phosphogluonate dehydrogenase gene) in two species of heliconiid butterfly showed that males produced twice as much of this enzyme as did females (Johnson and Turner 1979). This is in stark contrast to the common situation in both mammals, in which one copy of the homogametic chromosome pair (XX) is inactivated in diploid cells, and fruit flies where X-linked genes in the heterogametic sex produce twice the amount of those in the homogametic sex. This does not mean that there is no sex chromosomal inactivation in female heterogametic species. Indeed, there is considerable evidence from a variety of species of Lepidoptera and from birds that one sex chromosome is inactivated, but it is the W.

The Consequences of Female Heterogametry

Little is known about the precise mechanism of sex determination in ZW systems. This is unfortunate because the alteration from male to female heterogametry may have had considerable effects on some evolutionary systems within these groups. Here are two teasing examples. Birds and butterflies are perhaps the two groups within which sexual dimorphism, involving bright male adornments, is most pronounced. Is it purely coincidental that these groups should have females with only a single copy of their Z chromosome? It may be, but it is also possible that this distinction is crucial. In many of these species, male adornments are thought to have evolved, at least in part, as a result of sexual selection by female choice. For this to be the case, there must be a genetic basis to female choice. Should the genes that determine female mating preferences be on the nonpairing part of the larger sex chromosome (X or Z), they are more likely to be expressed when they first arise by mutation if they are in a species in which females are heterogametic (i.e., have only one of the larger sex chromosome). Most novel mutations are recessive, which is to say they are not expressed if there is a second, dominant allele present. Thus, a new recessive mutation that arises and gives its female bearer a preference to mate with males having a particular adornment will only be immediately expressed if the female is ZW (or ZO). In species with XX females, the alternative allele on the other X chromosome will block the expression of the new recessive mutation.

An additional advantage to a preference gene on the W chromosome arises if a sex-limited male trait, which females choose, develops into a handicap through Fisherian runaway sexual selection. As the preference gene never occurs in the sons of preferring females, the preference gene never becomes associated with the handicap trait (Hastings 1994).

A second intriguing observation involves the lack of chromosomal crossing over during meiosis in one sex of some organisms. It is fairly well known that there is no chiasmata formation, that is, no crossing over, in male *Drosophila*. Rather less well known is the fact that there appears to be no chiasmata formation in one sex in the Lepidoptera, but here it is females that lack crossing over. Whether it is merely coincidental that in both groups chiasmata formation is suppressed in the heterogametic sex is not known, but I am always somewhat suspicious of apparent biological coincidences. As yet, we understand too little about the intricate behavior of meiotic chromosomes to explain the method of chiasmata suppression. However, a detailed comparison of the ultrastructure and molecular chemistry of drosophilid and lepidopteran chromosomes during the four-strand pachytene stage of meiosis (when crossing over normally occurs) would certainly deserve attention.

Males without a Y Chromosome

The pattern of the sex chromosomes in which females have two similar Xs and males have one X plus a second, usually smaller sex chromosome, the Y, is seen in the majority, but by no means all, of sexually producing organisms that have true sexes. A variety of other systems that control sex determination have evolved. Some are based on other sex chromosome arrangements, some on as few as a single gene, some on nongenetic factors, and, as we shall see, in a few cases sex determination involves other organisms.

Considering other patterns of sex chromosomes, perhaps the smallest adjustment occurs in those species, such as some grasshoppers, in which the Y chromosome has been lost. These species are usually described as XO species, females having two X chromosomes and males just a single X. Males thus have one less chromosome in total than females in these species. Males are thus still heterogametic, as they produce two types of gamete, half the sperm having one more chromosome (the X) than the other half.

In XO species, the absence of a Y chromosome gives an immediate and obvious answer to the question of whether it is the number of X chromosomes, or the presence or absence of the Y, that is important in sex determination.

As an example, we can consider the nematode worm, *Caenorhabditis elegans*, in which sex determination shows some similarities to the *Drosophila* system, despite the absence of a Y chromosome (Parkhurst and Meneely 1994). Normal worms of this species are either XO and male, or XX and hermaphrodite. The system again depends on a complex of genes and a dosage compensation system. The differentiation of somatic tissue into male or hermaphrodite soma involves a cascade of expression of some ten genes. This cascade, as in *Drosophila*, is determined by the X:A ratio, although in *C. elegans*, the method by which this ratio is estimated is not clear. Dosage compensation in *C. elegans* also depends on the X:A ratio.

XO is a fairly common sex chromosome composition, a few XO (or ZO) species occurring in a wide range of different taxonomic groups.

Evolution of the Sex Chromosomes

Sex chromosomes have undoubtedly evolved independently many times in different taxonomic lineages. There are a variety of taxa, including reptiles, amphibians, fish, insects, and crustaceans, that contain some species that are male heterogametic (XY), others that are female heterogametic (ZW), and still others that are XO or ZO. In other groups, only one system occurs. Thus, in birds, all species are female heterogametic while all mammals exhibit male heterogametry. There is no homology between the ZW chromosomes of birds and the XY chromosomes of mammals, showing that the sex chromosomes in these groups evolved independently.

Because we know that the sex chromosomes of birds and mammals evolved independently, even though these two lineages had a common ancestor, we can consider the evolution of sex chromosomes in these groups. The model proposed is likely to be pertinent to the evolution of sex chromosomes in other taxa where the smaller sex chromosome plays a critical part in sex determination. The probable sequence of events is that one chromosome evolves a sex determination function and that recombination between this chromosome, which will become the Y (or W) chromosome, and its homologue is suppressed. Thereafter, chromosomal rearrangements that bring genes that are beneficial to the heterogametic sex, but not to the homogametic sex, onto this sex-determining chromosome will be maintained. The lack of crossing over prevents the linkage created from being broken. At the same time, other genes on this chromosome will degrade because of the accumulation of deleterious mutations in the absence of crossing over. Therefore, the sex-determining chromosome and its homologue will continue to diverge.

At one time it was thought that the Y and W chromosomes were almost completely lacking in genes. However, recent research in mammals and birds has shown this not to be the case. While genes sparsely populate these chromosomes, when they have a critical sex-determining function, the genes that they do contain have considerable influence. In mammals, many of the genes that occur on the Y chromosome are expressed only in the testis. These genes appear to influence sperm formation, as might be expected. However, some of the Y chromosome genes also influence sexually selected male traits such as body size and tooth development. These findings bear out a prediction made as early as 1931 by Fisher. Fisher realized that a mutant gene that was advantageous to the heterogametic sex but disadvantageous to the homogametic sex would be much more likely to spread through a population if it were linked to the sex-determining gene(s) on the Y (or W) chromosome because it would then only occur in the sex in which it was advantageous (Fisher 1931).

In an excellent review of this fascinating subject, Roldan and Gomendio (1999) comment that future research would benefit greatly from clear predictions as to which genes are likely to be found on the Y chromosome. It would then be my prediction that in birds and in those Lepidoptera in which the W chromosome plays a critical role in sex determination, a significant proportion of the genes that occur on the W chromosome will be mate choice genes.

Evolution within the Sex Chromosomes

It is an essential aspect of Mendel's laws of inheritance that genes occur in pairs because each chromosome has a homologue. Homologous autosomes pair during meiosis, and may form chiasmata during the first stage of meiosis, with the result that new gene combinations arise through recombination. However, the sex chromosomes are different in this respect. Although the sex chromosomes pair up during meiosis, the sex chromosomes in the heterogametic sex are not fully homologous. In most species the X and Y (or Z and W) chromosomes are quite different. They are homologous over just a short region (called the pseudoautosomal region). Over the rest of their lengths they do not pair up and so cannot recombine.

This absence of recombination between X and Y (or Z and W) chromosomes has a variety of consequences for the evolution of genes on these chromosomes. Many of the consequences for the way evolution acts on these genes result from this lack of recombination and/or one of three other features of these chromosomes. First, genes on the nonpairing region of the Y chromosome are only present in males. Second, recessive genes on the nonpairing part of the X chromosome are expressed when they occur in males because males cannot carry a dominant allele of the same gene because they only carry one copy of such genes. Third, in a population with a 1:1 sex ratio, there will be three times as many X chromosomes as Y chromosomes. As the Y chromosome has a lower population size, it will be more susceptible to the effects of random genetic drift, which generally acts to reduce variation.

So what are these consequences? The first involves the accumulation of deleterious mutations. Because the Y chromosome does not recombine, it acts essentially like an asexual haploid. Thus, it will be prone to the same type of accumulation of mildly deleterious mutations that was described previously when considering the advantage of sex through Muller's ratchet (p. 32). It has been argued that to avoid this possibility, selection may favor a reduction in the size of the Y chromosome. Indeed, in many species, the Y chromosome carries rather few genes and those that it does carry tend to code for traits involved in sex determination, male sexual function, or sexually selected male secondary sexual characters that would be costly to females if expressed in females.

The arguments for the degeneration of the Y chromosome follow directly from the proposal that the Y chromosome is essentially asexual and will be prone to the accumulation of mutations through Muller's ratchet. Brian Charlesworth (1978, 1991), who first proposed this idea, argues that the mutational load that results will lead to selection in favor of any translocations of functional genes to the X chromosome or the autosomes, as long as they are not specifically for male-determining traits or sexually selected male traits that would be costly in females. The Y chromosome should thus gradually decline in size, retaining only those functional genes that are advantageous specifically to males.

Simulations by Rice (1987, 1992, 1994), have provided empirical evidence in support of Charlesworth's theory. Using captive populations of the fruit fly, *Drosophila melanogaster*, Rice showed that sexually antagonistic genes (those beneficial to one sex but costly to the other sex) become associated with the sex-determining genes of the sex in which they are advantageous. He also demonstrated that in the absence of recombination, mutations do build up remarkably quickly. Observations of another fruit fly, *Drosophila miranda*, by Steinemann and Steinemann (1992) allow speculation of a molecular mechanism for degeneration. The Y chromosome of *D. miranda* consists of two parts: a section that shows homology with Y chromosomes of closely related *Drosophila* and a section that shows homology on part of chromosome 3 of other drosophilids and is assumed to have translocated to the Y chromosome in *D. miranda*. The translocated section is now in the process of degeneration due to transposable elements jumping into functional genes on this section, causing them to become nonfunctional.

An additional consequence of the absence of recombination between X and Y chromosomes is that genes that produce the phenomenon known as meiotic drive, which entails an overrepresentation of these genes in the products of meiosis, may be more likely to arise and be maintained (see chapter 4). Sex-linked meiotic drive genes promote passage of their own chromosome into progeny by killing gametes that carry their homologue. They are protected from destruction by a linked gene that makes them insensitive to their own action. If recombination could occur between the drive gene, and the insensitive gene a suicidal chromosome could be formed (Haig and Grafen 1991).

Single-Gene Sex Determination

Why are sex chromosomes of different sizes so common among animals? The evolution of anisogamy, which, in effect, is the evolution of sexes as males and females are defined in terms of differences in gamete size, must lead to some genetic control over the production of the sexes. Given the selective advantage of the 1:1 sex ratio, genetic control allowing the equal production of males and females would also be selectively favored. Perhaps the simplest way in which both might be achieved would be if sex were determined simply by one gene with two alleles. If one sex is heterozygous and the other homozygous, a 1:1 segregation of the sexes would automatically result. This type of system does exist in some species, such as the housefly, *Musca domestica.* However, rather few species adopt this simple method of sex determination. The reason for this is possibly that single-gene sex determination may be prone to corruption by inherited microorganisms or other genetic elements that favor the production of one sex or the other due to their mode of transmission (see chapter 4).

Sex Determination in Aphids: Two Homogametic Sexes!

In most of the species we have considered so far, reproduction is primarily sexual. However, in some species, a sexual generation alternates with one or more asexual generations in which just females are produced. In aphids, for example, females appear to be the homogametic sex having two X chromosomes. The question here is this: How do female aphids switch from production of XX females to the production of males with a different sex chromosome complement? Obviously, since the females that produce males have no Y chromosome and they do not mate, the sons that they produce cannot have a Y chromosome either. Indeed, male aphids are XO. Females can produce males by a modified form of cell division. In most aphids, parthenogenetic females produce further females by apomixis; that is, the reduction division of meiosis is suppressed. Males are produced in the same way, but the X chromosomes behave in an unusual manner. Rather than separating to move to the opposite poles of the two daughter cells, the X chromosomes remain behind in the middle of the meiotic spindle when the autosomes migrate. The X chromosomes then divide, but only one X chromosome migrates into the egg, the other X passing into the polar body. These eggs thus have only one X chromosome, but the normal diploid complement of autosomes. In effect, then, XO males arise because the X chromosomes undergo a kind of meiotic reduction division of their own (see Blackman 1974 for review).

The manner by which sperm are produced by male aphids is also unusual.

Because all the products of mating between male and female aphids are female, the sperm produced by males must all be the same with respect to the sex chromosomes. Spermatogenesis is by meiosis. At anaphase in the first meiotic division, the single X chromosome becomes stretched between the two poles of the dividing cell. It is as though the two daughter nuclei are having a tug-of-war over which should get the X. Then, as the two begin the final process of division, with the cleavage furrow starting to close between the two new nuclei, the X migrates into one of the daughter cells without dividing. This cell divides again in the normal second phase of meiosis to produce two haploid sperm, both of which have an X chromosome. The other daughter cell, which did not receive an X, simply degenerates.

Here, then, we have a very peculiar situation. The sex that has two X chromosomes, females, is homogametic because when it produces haploid gametes, these all carry an X chromosome. However, the other sex, the males, is also homogametic, for although males are XO, they also only produce sperm that all carry an X chromosome (Figure 3.4).

The factors that cause females to produce winged rather than apterous females and sexuals rather than parthenogenetic individuals vary between species, but crowding, host plant nutritional quality, nonnutritional characteristics of the host plant, temperature, and day length have all been shown to have an influence. In some species, one factor is of overriding importance. Thus, for example, in the vetch aphid, *Megoura viciae*, winged females are only produced if crowding causes the aphids to frequently come into contact with one another. Conversely, in many species, several factors are involved. Dixon (1998) has argued that by responding to multiple factors, aphids are able to track changes in the environment more closely, with the result that reproduction can be optimized.

Haplo-Diploidy

Sexually reproducing haplo-diploid organisms have haploid males and diploid females. Males are produced most commonly, but not exclusively, from female gametes that develop despite not being fertilized. Thus, males are the products of parthenogenetic reproduction. Females are usually the products of normal fertilization. Haplo-diploidy occurs in a diverse range of insect orders (e.g., Hymenoptera, Thysanoptera, Hemiptera, and Coleoptera) and in some other invertebrates, such as mites, ticks, and rotifers. Although this list of groups in which males are haploid and females are diploid appears small, approximately 20% of all animal species have haploid males.

There are two main mechanisms for the production of haploid males but diploid females. First, in some insects and mites, all progeny are produced from fertilized eggs, but the paternal sex chromosomes are ejected from eggs that are to become males (Bull 1983). Much more commonly, arrhenotokous reproduction is combined with haplo-diploidy. Males result from unfertilized and so

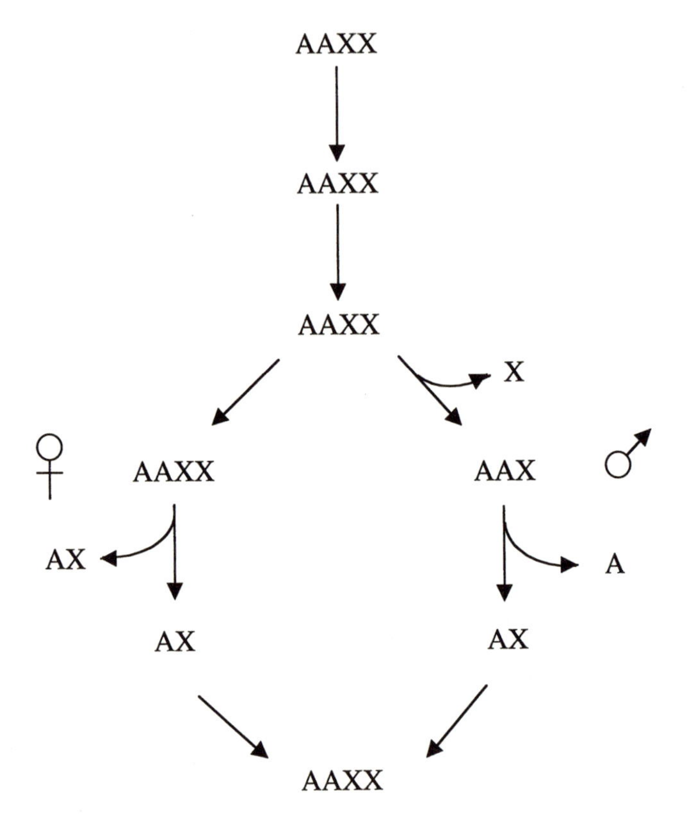

Figure 3.4 The sex chromosome cycle of aphids. Males and females are produced parthenogenetically, and usually viviparously. Sex determination occurs in the mother. Sexual females are then produced without meiosis. Males are produced by a modified meiosis in which only the sex chromosomes undergo a reduction division, one X being discarded. Finally, both females and males produce entirely haploid, X-bearing gametes.

haploid eggs, while females are produced by normal sexual reproduction (Figure 3.5). This combination of haplo-diploidy and arrhenotoky has evolved independently about a dozen times (Bull 1983). It is the normal mode of reproduction in the Hymenoptera and it is in this order of insects that haplo-diplontic biology has been most scrutinized.

The evolution of haplo-diploidy requires three fundamental genetic changes. First, genes must arise that prevent some of the eggs of a female from being fertilized. Second, a mechanism that causes unfertilized haploid embryos to begin to develop must arise. Third, some mechanism that prevents the diploid zygotes from developing into males is needed.

The first two changes, which have to occur simultaneously and may be under

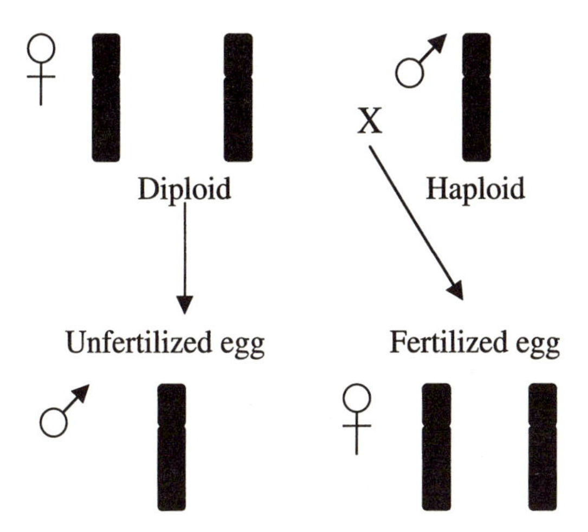

Figure 3.5 Sex determination in species in which haplo-diploidy is combined with arrhenotokous parthenogenesis. Females are produced from fertilized eggs and so diploid, while males derive from unfertilized eggs and are thereby haploid.

the control of the same genes, can confer a considerable advantage on the genes in question. Because haploid males transmit all their genes to the next generation in sperm, the genes in males are transmitted at twice the rate of genes in females that reproduce sexually, for they only transfer half of their genes in each gamete. The sexual progeny of these females contain half-maternal and half-paternal genes. If the father of these progeny is diploid, only half of his genes are passed to each offspring. However, if the father is haploid, all his genes are transmitted. This means that haploid-inducing genes will invade as long as the fitness contribution to offspring provided by haploid males is more than half of that which would be provided by diploid males. Once haplodiploidy has evolved, it is likely to be maintained because of the greater potential rate of transmission of paternal genes compared to maternal genes. Because females can also pass on genes through unfertilized eggs, the sex ratio produced by females now becomes critical. If females produce more haploid males than diploid females, there will be no advantage to the genes that induce haploid males. Indeed, these genes will now be at a selective disadvantage because the next generation will contain more maternal than paternal genes. However, the sex ratio of most Hymenoptera is strongly female biased, which means that at each generation a greater proportion of paternal genes are transmitted and the haploidy of males will be maintained.

The third change, the initiation of development of unfertilized eggs, I find extraordinary. Consider the implications of this change. In a normal diploid, females produce eggs. In some species at least, these can be retained in a mature condition for significant periods within the female, until they are fertilized. Fertilization is the catalyst for the beginning of zygote development. However, the haploid eggs of Hymenoptera that are to be male never receive this stimulus. Their development must start without it, but it must not start too early or one could envision males hatching while still within their mother. Perhaps the most sensible way to initiate development would be for the mother to provide some stimulus when she lays her eggs. Remarkably, this is exactly what has been observed in a variety of Hymenoptera. Male embryonic development is catalyzed by mechanical squeezing as the egg is passed through the narrow ovipositor. In species, such as the honeybee, *Apis mellifera*, in which the ovipositor has lost its egg-laying function and has been converted into a stinger, eggs are laid through the ovipositional pore. This procedure does not mechanically stress the eggs. However, the development of mature haploid eggs from honeybee queens is also activated by mechanical pressure. Eggs are squeezed through a narrow passage of strong circular muscle between the central oviduct and the vagina on the way to the ovipositional pore (Sasaki et al. 1996). In a wonderful series of experiments, Ken Sasaki and Yoshiaki Obara have also shown that either fertilization or mechanical stress will cause diploid eggs of honeybees to begin to develop. In this instance, then, embryo development seems to be initiated twice.

Degrees of Relatedness in Haplo-Diploids

The evolution of haplo-diploidy has a number of interesting effects from an evolutionary genetic perspective. First, as male gametes are produced without meiosis, all these gametes are genetically identical. This means that the degree of genetic relatedness in haplo-diploids is different from that in normal diploids. As all female offspring of a mating between a female and male obtain the same genes from their father and half the same genes from their mother, who produces her gametes through normal meiosis, these sisters will have not half of their genes in common, but three-quarters. This means that in haplo-diploids, full sisters (same mother and same father) are more closely related to each other than they are to either of their parents, or indeed than full sisters are to one another or to their parents in a normal diploid species. The genetic contribution of a queen hymenopteran to a daughter or son is 50%: they contain half her genes. The unusually high genetic relatedness of hymenopteran full sisters means that a female will benefit more copies of her own genes if she rears a full sister than she would by reproducing herself.

The high level of relatedness of full sisters can be contrasted to the degree of relatedness between brother and sister. Calculating the degree of relatedness between brother and sister depends on which way around one looks. From the male point of view, half of a male's genes are carried by his sister. All these have come from the mother. However, only a quarter of the sister's genes are carried by her brother because he gets none of the genes that she received from her father (Table 3.4). This difference in relatedness between full sisters compared to brothers and sisters explains why it is only female social Hymenoptera that become sterile workers. By helping to rear full sisters, they aid the passage

TABLE 3.4
Degrees of genetic relatedness in haplo-diploids. The relatedness, given as the relatedness of individual X to individual Y, is shown from both the male and female perspective. (After Crozier and Pamilo 1996.)

	Individual X	
Individual Y	*Female*	*Male*
Mother	1/2	1/2
Father	1	0
Daughter	1/2	1/2
Son	1	0
Sister	3/4	1/4
Brother	1/2	1/2
Aunt	3/8	3/8
Niece	3/8	1/8
Nephew	3/4	1/4

of their own genes more than they would by becoming sexual reproductives. This situation can be contrasted with another group of social insects, the termites. Termites are normal diploids and the relatedness between full sisters is just a half, as it is between all termite full siblings. Notably, in termites, both females and males may develop into sterile workers.

Calculations of the relatedness of male and female members of a colony or nest of social Hymenoptera are not always as straightforward as the above might suggest. In the discussion so far, I have assumed that all females in a colony are full sisters, which is to say the colony has just a single queen and she has only mated with a single male. The more sexually reproducing founders there are for a colony, the less closely related the members of that colony will be. So, for example, if a single queen who has mated with two different males founds a colony, then sisters will on average share half their genes. This is the same as in diploids. However, in a colony founded by a group of sisters, as occurs in bees of the genus *Trigona*, helping to rear female offspring is still more likely to evolve in haplo-diploid than diploid species. This is because the relatedness between cousins is greater in the former (37.5% of genes in common) than the latter (25% of genes in common) (Table 3.4).

The high relatedness of sisters also puts sterile workers in conflict with their queen. Workers gain more genetic benefit by helping full sisters than by helping to raise brothers. Thus, they benefit if the sex ratio produced by their queen is female biased. Conversely, the queen gains as much from producing sons as daughters because both contain half her genes. The different interests of queens and sterile workers over the sex ratio produced in a colony has led to what may be termed "intracolony conflicts" over sex allocation. The conflict may be put very simply. Because a queen gains as much from sons as from daughters, she will fertilize just half her eggs, thereby producing equal numbers of male and female eggs. However, because the workers are three times as related to the female eggs (assuming full sisterhood), as to the male eggs, if resources are limited the workers should give these to the immature females rather than the immature males.

This antagonism between sexually reproductive and sexually sterile females has been the subject of a great deal of theoretical and experimental research over the last 25 years. In fact, the 3:1 ratio in the relatedness of workers to full sisters compared to full brothers leads to the prediction that workers should care three times as much for immature females as for immature males. In an attempt to test this hypothesis, Bob Trivers and Hope Hare looked at the ratio of investment into male and female progeny in 21 species of ants in which colonies result from one singly mated queen. The ratio of investment they found was not quite 3:1 in favor of females, but was much closer to this ratio than to the 1:1 ratio that would be expected if sex allocation were controlled by queens. Trivers and Hare (1976) concluded that the workers won the conflict, largely because they could determine the sex ratio of the offspring that they raised.

An alternative hypothesis, that of local mate competition (p. 43), could mean that a female-biased sex ratio was also in the queen's interests. It thus seems that additional details of the biology of social Hymenoptera have to be known to successfully determine whether workers or queens have supremacy over sex ratios in these insects. In a small but growing number of cases, the requisite details have been obtained. The result is that in some species, the workers win; in others, queens win; and in still others power is shared.

In a thorough analysis of theory relating to sex ratios, specifically in ants, this is the conclusion reached by Andrew Bourke and Nigel Franks (1995). Summarizing their review, they note that although many factors may play a part in the allocation of investment into male and female progeny in ants, the various contributing elements can be incorporated into a unified theory based on Fisher (1930) with modifications by Trivers and Hare (1976). This unified theory allows predictions of sex ratios to be made for particular social and genetic scenarios. Considering these predictions in light of appropriate empirical data, Bourke and Franks (1995) conclude that, overall, in ants, sex ratio data support the concept of worker control proposed by the Trivers/Hare hypothesis, and the application of this theory to a variety of different social and genetic situations. They also note, however, that in some situations, other factors override worker control, although local mate competition rarely has an influence in ants.

Another consequence of the haplo-diploid system concerns the reproductive values of the sexes. From the above discussion, it is already obvious that females contribute more to the gene pool of the next generation than do males: half the genes of females and all the genes in the males of the next generation derive from the female parent. The higher reproductive value of females will lead to selection acting differently on males and females.

A final consequence of the haplo-diploid system is that inbreeding depression is likely to be low because males are haploid. Inbreeding depression results from bringing together in the same genotype deleterious recessive genes so that they are expressed. The frequency of such genes can rise to appreciable levels through random genetic drift simply because the recessives are not expressed in heterozygotes where their effect is suppressed by the presence of the dominant allele. As the recessives are not expressed in heterozygotes, they are not selected against. Of course, since male Hymenoptera are haploid, deleterious recessives have nowhere to hide. They will be expressed and so exposed to selection whenever they occur in males. Given the low level of inbreeding depression in the Hymenoptera, mating between close relatives should pose no problem. This leaves the way clear for distortions in the sex ratio as one of the assumptions of Fisher's sex ratio theory—that mating involves outbreeding—is contravened.

Haplo-Diploid Sex Determination

Setting aside for a moment the consequences of haplo-diploidy, we need to consider the precise method of sex determination in haplo-diploids. It is all very well saying that females result from fertilized eggs and males from unfertilized eggs, but why should it be this way around? Why, for example, could diploid zygotes not develop into males? At the outset, we can dismiss the two sex determination systems based on sex chromosomes that we have discussed previously. Because there are no sex chromosomes as such, there is no possibility that one or another of the sex chromosomes plays a critical role in sex determination. Furthermore, since all males have one of every chromosome and all females have two of every chromosome, there is no chance of dosage compensation.

In fact, in many hymenopterans, diploid males are produced, but they are sterile. Analysis of the differences between diploid males and females has led to the finding of a single multiple allelic gene that determines sex. If zygotes are heterozygous for this gene, they develop into females. If they are homozygous they develop into sterile males.

Diploid Male Hymenoptera

The production of diploid males in many hymenopterans has also attracted much recent attention because of the impact that such production may have on the population dynamics of a species. Species from which diploid males have been reported are spread widely throughout the Hymenoptera, examples being known among the Symphyta (sawflies and allies), Aculeata (bees, wasps, and ants), and the Parasitica (parasitoid wasps). The majority of the 30-plus species from which diploid males have been reported have been identified in the last 30 years, and it is certain that the list of such species will continue to grow.

Diploid males in the Hymenoptera are sterile. They arise due to a system called *complementary sex determination* (CSD). The production of such males is a consequence of genetic factors acting in fertilized eggs that when homozygous cause the diploid zygote to develop as a male rather than a female. In the first hymenopteran in which the control of diploid sex was analyzed, *Bracon hebetor*, a single locus with multiple alleles was found to be responsible. Females were all diploids with two different alleles present at this locus (e.g., *AiAj*), while males were either haploid (e.g., *Ai* or *Aj*) or diploid and homozygous for the A gene (e.g., *AiAi* or *AjAj*). There is an important difference between this type of single-locus sex determination and other examples mentioned previously. In those other examples, each allele has a particular sex tendency: it promotes development into either a male or a female. Here, however, the sex determined is simply a result of heterozygosity. If only a single type of *A* allele is present in a genotype, the result is a male, irrespective of whether the genotype is haploid or diploid. Only when two different *A* alleles are present is a female produced.

The reason that males are sterile is that they produce diploid rather than haploid sperm. The production of these sterile diploid males is costly for the population, and selection should either act to eradicate this system of CSD or minimize its effects. The effects of CSD will be minimized if the number of alleles at the *A* locus increases and/or if inbreeding is reduced or prevented. Thus, the production of diploid males can have a strong influence on the population biology and mating systems of species in which it occurs.

The number of alleles at the *A* locus is a crucial determinant of the level of diploid male production. If there are just two alleles at equal frequencies, then, assuming random mating, half the diploid progeny produced will be homozygotes and so sterile males (see Box 3.1a). Under the same conditions, with three alleles at equal frequencies, the proportion of homozygotes produced drops to a third (Box 3.1b). With four it drops to a quarter and so on. The number of alleles of the sex locus has been estimated in a number of studies. In most cases, the number of alleles is between 9 and 20, so that the proportion of diploid males produced ranges from 1/9th (11.1%) to 1/20th (5%). At these levels, the wastage due to the production of sterile diploid males must still be considered significant. However, in the bumblebee, *Bombus terrestris*, and the fire ant, *Solenopsis invicta*, much higher estimates of 46 and 86 alleles, respectively, have been obtained. Interestingly, the high number of alleles in *S. invicta* comes from a study of this species within its natural range in Argentina. However, estimates from introduced populations in the United States were much lower (10–13 alleles). This suggests that much of the genetic variation is lost during introduction, presumably because many of the *A* alleles are not present in the founding population. A number of studies on several species have shown a general pattern that the level of diploid male production is higher in introduced, small, or isolated populations, presumably as a result of a reduction in the number of *A* alleles during population foundation or bottlenecks.

If the consequences of the different types of mating that can occur in species with CSD are considered (Box 3.2), it appears that selection may promote three types of mating preference: for haploid males, for partners with specific genotypes with respect to their sex alleles, and for nonsibs (Cook and Crozier 1995). There is little empirical evidence for any of these types of mating prefer-

Box 3.1 The proportion of diploid progeny that will be homozygous for the sex-determining gene, and so male, in Hymenoptera showing complementary sex determination (CSD). This proportion increases as the number of alleles of the sex-determining gene increases. a) When sex-determining gene has two alleles. b) When sex-determining allele has three or more alleles. (Based on Cook and Crozier 1995.)

a)

Parents	male	×	female
	A1 or *A2*		*A1A2*
Diploid progeny	*A1A1* or *A2A2*	+	*A1A2* or *A2A1*
	sterile males		normal females

b)

Three alleles, *A1*, *A2*, *A3*, respectively, with frequencies p = q = r = 0.333.

Diploid genotypes are then produced at frequencies:

		A1	*A2*	*A3*
		p	q	r
A1	p	*A1A1* (p^2)	*A1A2* (pq)	*A1A3* (pr)
A2	q	*A2A1* (qp)	*A2A2* (q^2)	*A2A3* (qr)
A3	r	*A3A1* (rp)	*A3A2* (rq)	*A3A3* (r^2)

Total homozygotes $= p^2 + q^2 + r^2 = 3 \times 0.333^2 = 0.333$

Total heterozygotes $= 2pq + 2pr + 2qr = 3 \times 2 \times 0.333^2 = 0.667$

Therefore, one third of diploid progeny are homozygous.

This can be extended to any number of alleles, with the proportion of diploids being one over the number of alleles in the population, assuming each occurs at the same frequency.

ence and laboratory and field tests of mating preferences, in a variety of monandrous and polyandrous species would certainly be valuable.

The intricacies of the consequences of haplo-diploidy for sex allocation, sex ratios and sexual selection, particularly in the Hymenoptera, are outside the scope of this book. For those with a special interest in this order of insects, reference may be made to Ross Crozier and Pekka Pamilo's (1996) book *Evolution of Social Insect Colonies: Sex Allocation and Kin Selection* and to Andrew Bourke and Nigel Franks's (1995) book *Social Evolution in Ants.* Although

Box 3.2 Types of mating possible in Hymenoptera exhibiting complementary sex determination (CSD) controlled by a single locus, *A*. (Based on Cook and Crozier 1995.)

Unmatched matings: A female *A1A2* mates with a haploid male carrying a different sex allele *A3*. All diploid offspring are thus heterozygous (half *A1A3,* half *A2A3*) and so are normal females.

Matched matings: A female *A1A2* mates with a haploid male that carries one of the sex alleles within the female's own genotype (*A1* or *A2*). Half of the diploid progeny then produced will be homozygous and so sterile males. The frequency of matched mating will depend on the number of *A* alleles in the population and their frequencies.

Diploid male matings: Females may mate with diploid males. These males produce diploid sperm that often fail to successfully fertilize the eggs. Such females thus only lay unfertilized eggs that develop into normal haploid males. The cost of producing just sons will depend on the population sex ratio and other elements of the species' reproductive system. Occasionally, diploid sperm do fertilize haploid female gametes, a triploid zygote being the result. However, these suffer from low viability, sterility, and the production of offspring lacking one or more chromosomes, so that the conclusion that diploid males are sterile, while being formally compromised in these instances, is, for practical purposes, still safe.

Inbreeding: Of course, inbreeding increases the probability of a matched mating. If siblings from unmatched parents (*A1A2* × *A3*) mate, there is a 50% chance that the male's sex allele, derived from his mother and thus *A1* or *A2*, will be the same as one of the alleles of his sister (*A1A3* or *A2A3*). This probability applies irrespective of the number of *A* alleles in the population or their frequencies.

some readers may find some of the algebra in both books a bit heavy going, these are both fascinating books on these extraordinary groups of insects.

Haplo-diploidy is not confined to the Hymenoptera, but occurs sporadically in some other orders of insects and in a number of other Arthropod groups and other phyla. Haplo-diploidy in the coccid mite, *Icerya purchasi*, produces one of the strangest sex ratios known, because pure females have never been recorded. Populations consist of a small number of haploid males and a very much larger number of hermaphrodites that contain within them both diploid ovaries and haploid testes. The hermaphrodites are thus a mosaic of diploid and haploid tissue. This exceptional trick is achieved during the initially completely diploid larval stage of those destined to be hermaphrodites. As the larva develops, a modified form of mitosis produces localized haploid nuclei. These divide, producing two small clusters of cells, which develop into the testes and are surrounded by the ovaries. Ovaries produce oocytes by a normal meiosis. These oocytes are usually fertilized by sperm from the same individual, although occasionally males can also mate with hermaphrodites and achieve some paternity. Mating between hermaphrodites does not occur. Oocytes that are not fertilized are laid with the fertilized eggs and develop into haploid males.

Environmental Sex Determination

In a few widely disparate taxa, sex is not determined by specific genes cueing a switch in development toward maleness or femaleness, but by environmental stimuli activating the switch. In species said to exhibit environmental sex determination, genes are still involved in the development of sexual characteristics and, indeed, all individuals of a species have all the genes needed to produce both a normal male and a normal female. However, which set of genes is switched on depends on environmental influences at certain critical stages of development.

The Effect of Temperature in Reptiles

In many reptiles, including certain turtles, tortoises, and crocodilians, the temperature at which eggs develop determines sex. In the American alligator, temperatures below 30°C between the 7th and 21st days of incubation produce all females, while temperatures in excess of 34°C result only in males. Crocodilians lay their eggs in large nests of rotting vegetation. The parents can manipulate the temperature to which eggs are exposed by moving the eggs around the nest: warmer due to the decomposing vegetation toward the center of the nest, cooler at the edge. In addition to the variation in temperature in different parts of a nest, incubation temperature may vary according to the habitat in which nests are built. American alligator nests built in wet marshes produce lower temperatures than those that are built in drier conditions. The result is that wet-marsh nests tend to produce almost exclusively female animals. For example, in Louisiana, where marsh nests are much more common than dry nests, the population sex ratio is approximately five females per male.

The temperatures required to produce males and females vary between species of crocodilians. So, for example, in Australian esturine crocodiles (Plate 1d), only temperatures within a degree or so of 31.6°C will give rise to males. Higher or lower temperatures result in mainly females. This method of sex determination has been used to their advantage by crocodile farmers. Male crocodiles grow much larger than females. For example, male esturine crocodiles in Australia grow to up to 8 meters in length. Females, however, rarely exceed 4 meters. Particularly when young, the males grow faster than females.

Crocodile farmers therefore maximize production by incubating eggs for meat and skin stock at precisely 31.6°C, so that virtually all production is of fast-growing males.

The question then has to be what selective advantage is there to sex being determined by nest temperature in reptiles? A variety of hypotheses have been put forward on this subject over the last 25 years and, more recently, some empirical evidence has been obtained in support of some of these hypotheses. Without going into detail, these hypotheses can be split into three types: those that contend that the mechanism is ancestral and has no adaptive significance; those that are based on group selectionist arguments; and those that propose that the mechanism has some adaptive significance to the individuals. Available evidence tends to support the adaptive role for temperature induced sex determination.

The adaptive hypotheses, of which there are currently six (Shine 1999), all involve some link between the temperature at which eggs are incubated and the differences in the life-time reproductive success of males and females. To give an example of the type of situation that might favor temperature-controlled sex determination, allow that temperature, in addition to affecting the sex of an embryo, also determines how quickly it develops and hatches from the egg. Then allow that hatching early in the season is advantageous, as it has been shown to be in many organisms from ladybirds to birds. Early hatchers may gain additional resources and grow larger than late hatchers. Allow then that the advantages provided by these additional resources are greater for females than for males. Thus, if lower temperatures early in the season produce females, as in American alligators, the optimum benefit is gained. So males that gain less advantage from early hatching will hatch later. Different patterns will be seen in different species, depending on which sex gains from early hatching.

Other adaptive models for the evolution of temperature-controlled sex determination have also received some support from recent experimental work (see Shine 1999 for review), but in reptiles this field is still in its infancy.

The Evolution of Environmental Sex Determination

In other groups, the reasons for the evolution of environmental sex determination are better understood. In some species, for example, fitness is size related. Where the environment offers a patchy distribution of nutrients and other resources that impact on the growth of an immature animal and the adult size that it attains, environmental heterogeneity will influence fitness. Imagine then that size variation has a greater influence on female fitness than on male fitness. Thus, while small males are only slightly less fit than large males, small females have very much lower fitness than large females. In such circumstances, immature individuals in low-resource environments that only allow restricted growth should develop into males, while those patches with plentiful resources, allowing potential growth to be fulfilled, should develop into females. Here then it makes evolutionary sense that the determination of sex is delayed and regulated by some factor in the environment that acts as an indicator of patch quality.

This relationship between size and fitness is precisely the case for some parasitic nematodes and crustaceans. Here, parasites in large, resource-rich hosts and hosts that have a low parasite-infection level develop into females. Small hosts or heavily infected hosts give rise to more male parasites. A similar rationale can be applied to some free-living organisms.

Neighbor Sex Effects

An individual's environment, of course, also includes other members of the same species. In some animals, conspecifics close by can have a strong influence on sex determination. *Boniella* is a marine worm that shows a high degree of sexual dimorphism, females being much larger and anatomically more complex than males. Females produce eggs that hatch into free-swimming larvae. These larvae settle on the sea bottom before becoming adult and it is only as they mature that their sex is determined. Those larvae that settle on or in close proximity to a female become male, while those that settle away from adult females themselves become female. Chemicals excreted by adult females engineer the switch. In the presence of these chemicals, larvae develop into males that then fertilize the females. In the absence of the chemicals, a settling larva develops into a female and begins to manufacture and excrete these chemicals, thus, in effect, making herself a mate.

Where Large Males Are Advantageous

A contrasting example to those detailed above involves species in which variance in fitness correlated to size is greater for males than females. This is the case for some fish (see below) and a number of crustaceans. In the brine shrimp, *Gammarus duebeni*, female fecundity is somewhat correlated to size. However, in males, variance in fitness is greater. This is because, during the reproductive season, males have to hold onto females to guard them from other males. A male's ability to hold onto a female depends on his size relative to hers. Large males can hold a wide range of females, while the smallest males fail to guard females of any size. Because *G. duebeni* size is largely a function of the length of time that immatures have for development in the year preceding reproduction, offspring born early in the breeding season (May—July), which have a long time to develop, will attain large size and so should develop as males. Those born later should develop as females. This pattern is borne out in *G. duebeni*, and laboratory experiments have shown that sex determination depends partly on photoperiod. Short day length tends to lead to the production of females (Bulnheim 1978) (Figure 3.6).

Sex Changers

Many species of fish that live in groups exhibit female-biased tertiary sex ratios. In some species, the female bias is the result of a capacity to change sex

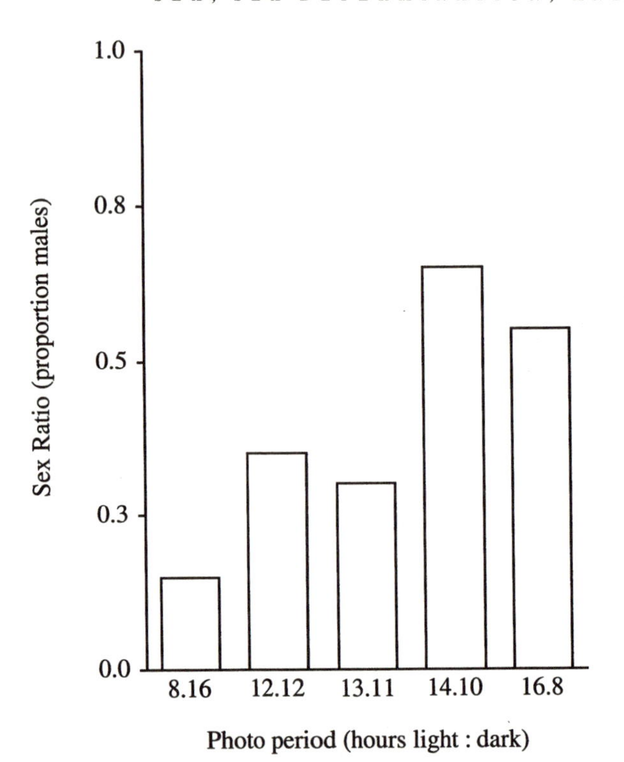

Figure 3.6 In the brine shrimp *Gammarus duebeni*, sex determination is partly due to day length. The histogram shows the proportion of males produced from the same brood when incubated at various day lengths. (Redrawn from Bulnheim 1978.)

even when adult. Gaimardi's wrasse, *Cardis gaimardi*, lives in small hierarchical groups of one silver-colored male and several females that are orange with blue spots. If the male of a group is killed, the highest-ranking female in the group begins to change color and sex, soon becoming male. Here all individuals are the same, initially developing as females and all having transsexual capacity if the occasion arises. A similar pattern is seen in the cleaner fish, *Labroides dimidiatus*. Males are territorial and control a harem of five or six smaller but mature females. Although the male is always larger than any of the females, the females in a harem develop a size-related dominance hierarchy. If the male dies, the largest female will take his place, controlling the harem and showing male behavioral traits within a few hours. Morphological transformation takes place rapidly and within two weeks the promoted individual can produce sperm.

The sex-change strategy in these cases may have evolved to reduce the risk of being a young male in competition with older and stronger males for possession of females. What better way of avoiding being beaten up by dominant males than being a potential mate rather than a rival? In fact, many fish species adopt this strategy of reproducing as a female when small and then transforming into a male when large enough to have a reasonably good chance in competition with other males. The strategy will be successful as long as the lifetime reproductive success of an individual that changes sex is greater than that of an individual that remains the same sex throughout its lifetime.

Of course, as is almost always the case in the natural world, the system is often more complex than this. In the case of the blue-headed wrasse, *Thalassoma bifasciatum*, an Atlantic coral reef fish, some fish are males from birth (primary males), yet females still have the capacity to change sex if the opportunity presents itself through the loss of dominant males. The reason for the existence of some fish that start and remain males throughout their lives is that these males adopt an alternative reproductive strategy. Rather than placing their reproductive faith in an ability to outcompete other males through size, these males are small and indulge in what is termed *sneaky mating*, infiltrating the territories of dominant males and attempting to spawn with one of the females that has been attracted to the reef master. On some occasions, bands of these small males will chase females, stimulating the females to spawn with the group in a form of external gang rape. Interestingly, primary males occur far more commonly in areas of high wrasse density, for here the opportunities for successful sneaky copulation are higher than on reefs with just a few wrasse, where the dominant male(s) can keep an eye on all the females and exclude sneaking completely (Warner and Hoffman 1980).

In only a few instances in vertebrates does sex change operate in the opposite direction, from male to female. In the clownfish, *Amphiprion akallopisos*, females and males form pair bonds, living in close harmony with sea anemones. Each pair has its own anemone. The resources available within the mutualistic symbiosis, in the form of space and protection provided to the fish by an anemone, are only really sufficient for a single pair of fish. The result is that in these fish, monogamy is imposed on them by their habitat. The success of a pair depends largely on the number of eggs produced by the female, so it is likely to benefit the pair if the larger fish is female. And this is almost invariably the case. If a female is removed from a pairing, a new fish will move in. The newcomer tends to be a small male, and then the old male, the larger of the two will transform into a female (Fricke and Fricke 1977).

Contrary to the situation in vertebrates, sex changes from male to female are much more common in invertebrates than the reverse. This largely results from a greater correlation between size and reproductive output in female as compared to male invertebrates. Increased size is more beneficial to females than males and among invertebrates females tend to be larger than males. Examples of feminization will be discussed in chapter 5.

Microbes That Change Host Sex

The final type of sex determination system is perhaps the strangest of all, for it involves the sex determination of one species being usurped by another species. In a few species of Crustacea and Lepidoptera, microorganisms that live in the cytoplasm of cells and are passed from mothers to offspring, but not from fathers to offspring, manipulate their hosts to increase the proportion of hosts that can transmit them. Because the inheritance of these microorganisms, which include both bacteria (*Wolbachia*) and a variety of protists (primitive eukaryotes such as Microsporidia), is exclusively maternal, the microbes convert genetic males into females so that they can be inherited. The result in some populations is that all individuals have the same sex chromosome complement, the only difference between males and females being that females are infected with the microbe while males are not, having lost it because not all of a carrier female's progeny inherit the symbiont.

This extraordinary situation arises ultimately out of the fundamental difference between females and males. Females produce large gametes, the nucleus being surrounded by a reservoir of cytoplasm, while the sperm of most species are little more than a nucleus with a tail. The cytoplasm of cells provides an environment in which microorganisms can live, and, in the case of female gametes, can be transmitted down the host generations. For such microbes, being in a female rather than a male host is crucial.

Although the number of species in which microorganisms are known to change the sex of their hosts is not large (details of these cases are given in chapter 5), this may be at least partly because relatively few species have been subjected to appropriate analysis. Because work over the last decade has indicated that a large proportion of invertebrate species harbor cytoplasmic symbionts, the number of known examples in which cytoplasmic microorganisms interfere directly with their host's sex determination may increase substantially in the future.

Parental Control of the Sex Ratio?

It is apparent from the foregoing descriptions that sex determination systems vary considerably between different types of organism. Some of the sex determination mechanisms described certainly should allow parental control of the numbers of sons and daughters that they produce. In particular, temperature-controlled sex determination in crocodilians, in which the parents tend the nest and can move eggs to positions where they are exposed to higher or lower temperatures, would allow parental manipulation. That these reptiles usually produce equal numbers of sons and daughters lends support to Fisher's theory for the stability of the 1:1 sex ratio.

In haplo-diploid species, where females have control over the proportion of eggs that they fertilize, and thus develop into females, and the number they do not fertilize, which become males, a variety of factors have been shown to influence sex ratio, Hamilton's local mate competition perhaps being the most important.

In other instances in which females appear to have some degree of unconscious control over the sex of their offspring—for example, female mammals producing one or the other sex depending on their own condition—the precise mechanism of control is not yet known.

Despite the potential for parental control over the sex ratio in a number of species, or possibly because of it, we are left with a set of hypotheses as to the optimum sex ratio in different circumstances (outlined in chapter 2) and a diverse array of sex-determining mechanisms. Selection should act to favor the sex determination mechanism that, within the genetic constraints of a species, allows for the optimum sex ratio predicted by the hypothesis that is most appropriate to that species to be achieved. In the main, this seems to be the case when an organism is left to its own devices. However, when the optimum sex ratio of the two sexes of a species, or different castes of a species, or different genes in a species, including genes carried by inherited parasites, differ, there is a conflict over the optimum sex ratio. In the next chapter, I will discuss some of these conflicts.

CHAPTER 4

Genetic Law Makers and Law Breakers

But this is an unending war
The war drum will forever throb
The battle flags will never be furled
The Parliament of genes will ever disagree
A federation the genome will never be.
—With apologies to Alfred Lord Tennyson

Summary

Most genes in sexually reproducing organisms are inherited equally from the two parents in the nuclei of the cells that fuse at fertilization. The pattern of inheritance of such genes is in accord with Mendel's laws of inheritance. However, some genes are not inherited equally from both parents, while others have mechanisms that increase their chances of inheritance compared to their allelomorphs. The aim of this chapter is twofold: to introduce some of the mechanisms employed by these genetic law breakers, and to explain how unequal patterns of inheritance lead to conflicts within the genome.

Examples of sex chromosome and autosomal meiotic drivers, selfish supernumery chromosomes, mitochondria that induce male sterility, and a variety of selfish strategies adopted by inherited intracellular microorganisms are introduced, with the phenomenon of cytoplasmic incompatibility being considered in some detail.

The question of why rather few genes or genetic elements flout Mendel's laws is considered at the end of the chapter.

Fair Genes and Foul

The greatest single step in genetics, the study of inheritance, was the discovery of the basic rules governing the passage of characteristics down through the generations. These rules, first discovered by Gregor Mendel, are called *Mendel's laws*. They describe how traits, controlled by genes carried by parents, segregate in their offspring or in the following generation and how each trait assorts independently. The basic patterns of inheritance, described by Mendel from his studies of pea plants, were soon extended to include genes on nonpairing sections of the sex chromosomes (sex linkage) and alleles of different gene loci that tend to be inherited together because they are situated close together on a chromosome (genetic linkage).

Mendel's laws, and the patterns of inheritance that they lead to, were combined with Darwin's theory of evolution through natural selection by a number of eminent geneticists (Fisher, Wright, Haldane, Muller, and others) in the first four decades of the twentieth century, leading to the new synthesis of evolutionary biology. The fields of theoretical population genetics and ecological genetics grew out of this neo-Darwinian synthesis (see for example Mayr and Provine 1980).

At the center of Mendelian genetics lies the theoretical 50-50 chance of any allele ending up in a particular gamete. The vast majority of studies of the inheritance of traits reflect this probability. However, in a few instances, traits have been observed to behave differently. As early as 1909, Correns observed that chloroplasts are inherited, but only from the male parent. Mitochondria were also shown to be inherited uniparentally in most species, although in this case, maternally. Other early anomalies were seen in the sex ratios produced in some *Drosophila* families (Morgan et al. 1925) and the abnormal progenic ratios with respect to the autosomal *t* locus in mice (Chesley and Dunn 1936). Such cases were viewed as oddities by most evolutionary biologists—exceptions that proved the rule.

Through the second half of the twentieth century, the frequency of discovery of oddities that defied Mendel's rules increased. Now, following the turn of the millennium, most weeks see the publication of another novel instance. The genetic elements that behave in this manner are associated with a range of different phenomena. In all cases, the non-Mendelian behavior of these elements is manifest in a transmission bias into the next generation that favors them. Indeed, this bias is at the root of their survival and success and can be explained by natural selection. Thus, these elements are not freaks of nature, but rather a normal part of the diversity of nature that have evolved through the same processes that have produced the leopard's spots, the giraffe's neck, the peacock's tail, and the human brain.

There are two methods by which a piece of DNA can increase its frequency through time. First, it can confer some advantage on its bearer. This was the mechanism envisioned by Darwin: inherited traits that increased fitness would be more likely to survive than those that decreased fitness would. Darwin did not have a clear understanding of how characteristics were passed from parent to offspring. However, following the rediscovery of Mendel's work in 1901, it became an implicit assumption that the two copies of an autosomal gene would be represented equally in the zygotes produced after fertilization. Selection acting from this time would bias the frequency of the alleles in the new generation in favor of the allele that conferred the greatest benefit. But what if some mechanism acts before fertilization, in apparent or actual contravention of equal Mendelian segregation?

Imagine a mutation in a gene that was expressed in sperm, allowing those bearing this allele to swim faster than sperm bearing alternative alleles. Sperm carrying this "fast" allele will tend to win the race to the eggs and so be over-represented among progeny. The "fast" mutation is likely to spread rapidly through the population, irrespective of whether progeny are fit in other ways. Because the number of progeny produced is usually a function of the number of eggs produced rather than the number of sperm, there will be no cost involved in terms of the number of zygotes formed. Professor J.B.S. Haldane (1932) applied this idea to plants, writing:

> Clearly a higher plant species is at the mercy of its pollen grains. A gene which greatly accelerates pollen tube growth will spread through a species even if it causes moderately disadvantageous changes in the adult plant. (p. 123)

In the sperm case, we can now change the effect of the mutation from fast swimming to assassin. If the mutation caused its bearer sperm to kill sperm that did not carry it, again it would spread, but now there might be a cost through reduced fertility. However, since most organisms produce sperm in considerable excess, this cost might be low or negligible.

If we change the effect of our selfish mutation again to cause not the death of sperm but the death of sibling embryos that do not carry it, the mutation will obviously have a cost in terms of the number of progeny produced by a parent. In this case, the mutation is not present in any more progeny than it would have been anyway, so the benefit it gains is not so obvious. However, the mutation may gain benefit in two ways—one relative and the other direct. First, because embryos carrying the nonassassin allele die, the relative frequency of the assassin allele will increase. Second, if the mother of a set of offspring nourishes the zygotes once formed, the death of half of them will lead to the remaining half, the assassin bearers, gaining more resources, thereby increasing their chances of survival.

From these theoretical examples, it is clear that scenarios exist which permit the spread of genes that promote their own continued existence, even if they reduce the fitness of their bearers. Genes that break the laws of Mendelian genetics have been given a variety of different names, such as outlaws, or ultraselfish genes. The critical aspect of the evolutionary spread of such genes is that selection acts directly on the genes (or the chromosomes, organelles, or symbiotic microbes that bear them) rather than on the individuals that bear them. In this chapter I shall describe a range of cases in which Mendel's laws are violated, consider the conflicting interests of these genetic outlaws and the rest of their bearers' genes, and discuss some of the mechanisms by which "law-abiding" genes protect themselves against these anarchistic genetic elements.

Conflict within the Genome

Most genes are inherited with equal likelihood from either parent. However, some genes are inherited mainly or only from one parent. These genes are important because, as a result of the bias in their transmission, they are in conflict with the majority of the genome that plays by Mendelian rules. Genes inherited maternally include genes on the nonpairing part of the W chromosome (in female heterogametic species), cytoplasmic genes whether they be on mitochondria or in intracellular symbionts, and a few extracellular symbionts. Paternally inherited genes in animals are almost exclusively confined to those on the Y chromosome in male heterogametic species.

In addition to genes that are inherited from just one parent, there are genes inherited from both parents that are not inherited equally from both parents. Here we are thinking about genes on the nonpairing part of the homogametic sex chromosome; that is, on X or Z chromosomes. Such genes are transmitted equally from both parents to progeny that are of the homogametic sex. Thus, girls receive one X chromosome from their mother and one from their father. However, genes of this type are only inherited by progeny of the heterogametic sex from their homogametic parent. Boys receive their X chromosome from their mother, not from their father. Put another way, two-thirds of the X chromosomes in a population are derived from female parents simply because females have two copies of this chromosome, while males have only one. In female heterogametic species, only a third of the Z chromosomes are derived from female parents, two-thirds being inherited from males.

It is in the interests of these genes to promote the sex through which they are most likely to be inherited. Thus, these sex-linked genes are in conflict with the rest of the genome over the number of each sex of progeny that should be produced. For law-abiding genes, a 1:1 sex ratio is usually stable (p. 39). However, selection acting on sex-linked genes would favor strategies that caused these genes to bias the sex ratio in favor of the sex through which they are transmitted. Observations of heritable distortions in the sex ratio are then usually seen as evidence of this conflict within the genome.

In most instances, sex ratio biases are in favor of females. This is simply a consequence of there being far more genes that are inherited most often or exclusively from female than from male parents. The mechanisms by which sex ratio distortion is achieved by these self-promoting genes, in those cases in which the mechanism has been clarified, are varied and imaginative. They include gametic assassination, sex change, abolition of sex, and assassination of both nonbearing and nontransmitting carriers or hosts.

Ultraselfish genes that increase their own transmission by distorting the sex ratio of their carriers can be split into three groups. These are nuclear genes on the chromosomes or parts of chromosomes that do not pair normally at meiosis; genes carried on cytoplasmic organelles of their host, such as the mitochondria; and genes of cytoplasmic symbionts. Considering the first of these groups, I will begin by looking at how some genes on sex chromosomes break the rules.

Driving Sex Chromosomes

Gametes are produced by the process of cell division known as meiosis. When an allele cheats during meiosis, so that it is present in more than half of the gametes formed, the allele is said to show meiotic drive (Sandler and Novitski

1958) or segregation distortion. If this allele is on the sex chromosome, this drive will cause a distortion in the sex ratio.

The earliest examples of meiotic drive were observed in three species of fruit fly (*Drosophila* spp.). In each case, some female flies were seen to produce more daughters than sons. Careful breeding experiments showed that daughters in the female-biased families produced equal numbers of males and females. However, although half of the sons of these females (i.e., the grandsons of the original pairs) produced a 1:1 sex ratio, the other half produced a female bias, irrespective of where the females came from.

Detailed analysis of the system showed that the bias was in the primary sex ratio, since no difference in viability could be found between eggs from the odd sex ratio (SR) pairs and normal (N) pairs. Furthermore, because SR was derived from some feature of males rather than females, parthenogenesis could be discounted as an explanation. The possibility of genetic males being feminized also seemed an unlikely explanation because such feminization is usually associated with the production of at least some intersexes, and none were observed in the crosses. The deduction, reached largely by exclusion of other possibilities, was that sperm from males carrying the X chromosome were more successful in achieving fertilization than those carrying the Y chromosome. This was either the result of a bias in the number of X-bearing sperm over Y-bearing sperm available, or because X-bearing sperm outcompeted Y-bearing sperm in the fertilization race.

The precise mechanism of drive in this case was revealed by electron-microscopic examination of the testes and sperm bundles of males in SR and N lines. SR males contained only about half the number of sperm that N males did. This reduction was attributed to mortality of the Y-bearing sperm. The pattern of a factor that acts in males, but is inherited through females, suggests that the death of Y-bearing sperm is due to a factor on the X chromosome. This makes intuitive sense, at least from the point of view of the X chromosome. By killing Y-bearing sperm, the Y-killer gene on the X chromosome increases its transmission into the next generation.

This example of X chromosome "drive" is thus a classic example of what Jim Crow (1988) called an ultraselfish gene: a gene that spreads despite, or, rather, as a result of, the damage it causes to its host.

Species that have been reported to show sex ratio distortion as a result of sex chromosome drive have been reported from widely different taxa. In addition to a variety of fruit fly species, these include mosquitoes, butterflies, moths, at least two mammals, the wood lemming, *Myopus schisticolor* (Fredga et al. 1977) and the varying lemming, *Dicrostonyx torquatus* (Gileva 1987), and one plant, the white campion, *Silene alba* (Hurst and Pomiankowski 1991; Lyttle 1991). However, recent research has cast doubt on some of the reported invertebrate instances outside the Diptera. In particular, in the Lepidoptera, all cases of female biases previously attributed to sex chromosome drive have been shown to be associated with maternally inherited symbionts that either feminize genetic males (p. 121) or cause the death of male, but not female, offspring (p.

157). Indeed, within the invertebrates, there is now no confirmed case of sex chromosome meiotic drive outside the Diptera (Jiggins et al. 1999a). Although it is possible that instances of sex chromosome meiotic drive will be discovered in other species of invertebrates, it does seem that the Diptera are a genuine "hot spot" for this type of non-Mendelian inheritance.

The reasons why Diptera are especially prone to this phenomenon are unknown. However, there may be a rational argument why sex chromosome drive is confined to species in which males are the heterogametic sex. If drive has to operate in the sex that has gametes containing two types of sex chromosome, as seems sensible, when males are heterogametic, the loss of about half the sperm is unlikely to have a major negative impact on fertility, because sperm is generally produced in huge excess. Conversely, were drive to operate in female heterogametic species, the loss of half a female's functional gametes, her eggs, might be strongly detrimental, because these are very costly to produce. Selection may therefore prevent the spread of a sex chromosome meiotic driver in groups, such as the Lepidoptera and birds, with heterogametic females. Various authors have argued that sex chromosome meiotic drive is important in certain fields of evolutionary biology, particularly those of sexual selection and speciation. However, if sex chromosome meiotic drive is largely confined to the Diptera and a few isolated species in other taxonomic groups, this seems unlikely.

Drive of Autosomes

Meiotic drive is not entirely confined to sex chromosome drive. There are a small number of reported cases of autosomal drive (two in mice, one in fruit flies, two in fungi). The paucity of reports of autosomal meiotic drive, compared to sex chromosome drive, may be a simple observation bias because the sex ratio distortion caused by the latter is rather obvious and easily detected. Since autosomal drive does not cause a distortion of the sex ratio, it need not concern us here for long. However, one finding from the study of autosomal drive is worth describing, for it gives an insight into how drive actually works and shows how the conflict involved in drive leads to an arms race.

Genetic analysis of an autosomal meiotic driver in the fruit fly *Drosophila melanogaster* has shown that here drive is the result of a single gene, *Sd*. This gene drives chromosome 2, so that in heterozygotes, the chromosome 2 carrying *Sd* is overrepresented in progeny. However, drive does not always occur. A second gene on chromosome 2, called *Responder* (*Res*), determines whether drive actually occurs (Figure 4.1). If the homologue of the *Sd*-carrying chromosome 2 carries the sensitive allele of *Responder* (Rsp^s), drive results and the *Sd*-carrying chromosome is transmitted in excess. However, if the homologue carries the insensitive allele, Rsp^i, no drive results. Notably, *Sd*-carrying chromosomes always carry Rsp^i. This is necessary because it prevents an *Sd*-carrying chromosome 2 from committing suicide, killed by the *Sd* gene it carries. Even more interesting is the fact that *Sd* and Res^i are inextricably linked, since they are situated within a piece of chromosome that has, at some stage in the

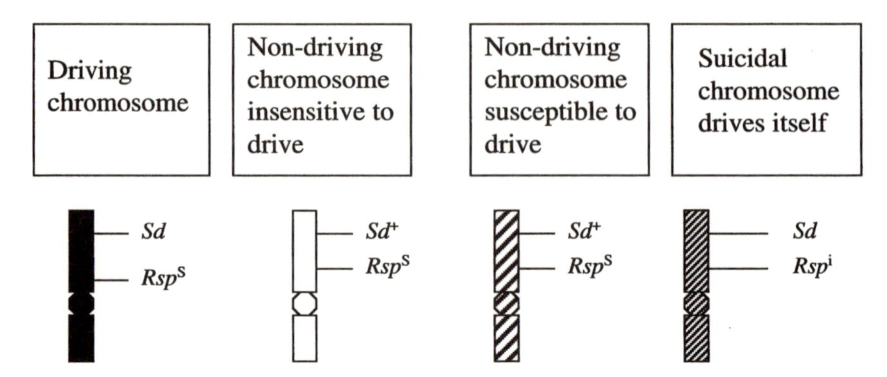

Figure 4.1 Autosomal drive of chromosome 2 of *Drosophila melanogaster*: Chromosomal types. (Redrawn from Majerus et al. 1996.)

past, broken out, flipped over, and gone back into the same hole. This flipped-over section is called an *inversion* and is unusual because it prevents crossing over. The driver allele is thus always inherited with its own insensitive allele and cannot be separated from it by meiotic crossover.

The Effect of Sex Chromosome Drive on Host Populations

It is easy to understand why driving sex chromosomes, once they arise, can spread through their host populations. Taking the case of drive in species with males being heterogametic (there is no confirmed case of sex chromosome drive in a female heterogametic species), the critical sex is the male. As males produce an abundance of sperm, there is scope for the sperm from one male to compete among themselves for fertilization of eggs. Thus, a sperm carrying a chromosome that causes the death of sperm not carrying this chromosome will have a greatly increased chance in the competition for fertilization. It will thus have an advantage over its homologue and will spread through the population at its expense. The dynamics of driving X and Y chromosomes, and their effect on population sex ratios, were first modeled by Bill Hamilton (1967) (Figure 4.2). Critical to our understand of the effect of driving sex chromosomes and their relative rarity, given that once they arise they should start to spread, is the prediction from Hamilton's model that the ultimate fate of a population invaded by such a chromosome may be extinction. This may occur if one sex becomes so rare that many individuals of the other sex (the sex favored by drive) fail to find a mate. However, one feature of the known driving sex chromosomes is that they are not totally efficient; that is, they do not eradicate their homologue entirely from the sperm that achieve fertilization. The consequence of this inefficiency is that an equilibrium level is achieved in the population. The precise equilibrium level achieved depends on the efficiency of the driver and the mating system, with the number of times that males can mate, if these are the rarer sex, having a strong influence.

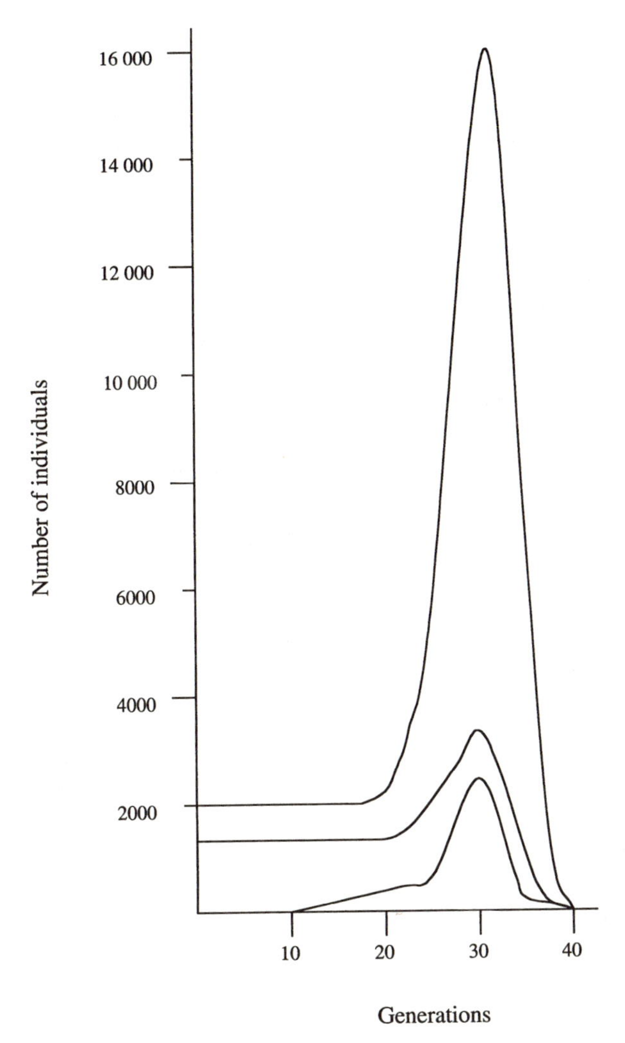

Figure 4.2 The dynamics of driving sex chromosomes. In this simulation, a driving X chromosome was introduced into a single male in a population of 2,000 individuals with a 1:1 sex ratio. Males can mate with a maximum of two females, and unmated females fail to reproduce. The lines give the total population size (top line), the number of males in the population (middle), and the number of males in the population that carry the driving X chromosome (bottom). (After Hamilton 1967.)

The distortion in the sex ratio produced by a driver will impose novel selection pressures on the population. One of these is easy to understand. As the population sex ratio becomes increasingly biased, there will be selection pressure on the rest of the genome to resist the driver by producing individuals of the rarer sex. This follows straight from Fisher's explanation of the evolutionary stability of the 1:1 sex ratio.

When one sex begins to predominate as a result of a sex chromosome driver, the rarer sex will have a greater reproductive potential. Therefore, mutations on any chromosome other than that being driven, which promote the production of the rarer sex, will be favored. The greater the sex ratio distortion produced by the driver, the stronger the selection in favor of the production of the rarer sex will be. The promotion of the rarer sex may be either by direct increase in the production of the rarer sex or through the suppression of the action of the driver. Consequently, we may expect drive resistance genes to spread until the sex ratio returns to parity or until an equilibrium is reached between the driver and the resistance genes.

There is good evidence that drive resistance genes, usually known as *restorer genes*, prevent drivers from causing the extinction of their host populations and are maintained in populations at equilibrium with drivers. In *Drosophila*, where the X chromosome is driven, both autosomal and Y chromosome restorer genes have been found. In the mosquito *Aedes aegypti*, the Y chromosome is the driver, and here Hickey and Craig (1966) have shown that X chromosomes vary in their susceptibility to drive.

One other question deserves consideration: Are all sex chromosome drivers inefficient when they arise? Currently we have no answer to this question. All we can say for the moment is that it is possible to imagine an evolutionary scenario in which the transmission advantage of a driver would diminish with time and this may lead to selection for a reduction in the driver's efficiency. Imagine a species with a driving X chromosome. The males of this species may mate many times, but produce a finite number of sperm per day. When males are still common, the driver has great success because it kills most of the sperm carrying a Y chromosome. However, as the driver spreads, males become increasingly rare. Eventually, males are so rare that although all females could mate with a male, the fertility of the males with each female starts to decline because they are becoming sperm exhausted. Males in which less of the Y-bearing sperm are killed by the driver will have an advantage (in addition to that resulting from producing some of the rarer sex) because they will have more surviving sperm and so be able to achieve higher fertility with the females that they mate with. As yet we have no empirical evidence to assess whether selection does lead to a reduction in driver transmission efficiency. What we can say is that it is probable from Hamilton's deductions that we are far more likely to find inefficient drivers simply because efficient drivers are likely to drive their host populations to extinction in a fairly short period of time.

B-Chromosome Drive

Chromosomes that are essential to the functional well-being of diploid organisms usually come in pairs. Even the sex chromosomes pair up during meiosis, although not along their whole length. However, many species of animals and plants harbor extra chromosomes that are not necessary to the life of their bearers and do not pair up in the normal way at meiosis. These are called *supernumerary* or B chromosomes to differentiate them from those required by the organism (A chromosomes). Some of these chromosomes persist in hosts because, while not being essential to the survival or reproduction of the individuals that bear them, they do increase fitness. For example, a B chromosome increases seed germination in the chive, *Allium shoenoprasum*. Other B chromosomes break Mendel's laws because they are driving chromosomes. Their spread and persistence depends on their being included in more than 50% of gametes.

Most driving B chromosomes do not affect the sex ratios of their hosts. However, the parasitoid jewel wasp, *Nasonia vitripennis*, bears a B chromosome that drives its own transmission in a most unusual way and does influence the sex ratio of its host (Werren et al. 1981). In this case the sex ratio distortion is in favor of males and the sex that bears the driving B chromosome is also male. The phenomenon is referred to by the name *paternal sex ratio* (PSR) because it is derived from fathers. *Nasonia vitripennis* is a typical haplo-diploid member of the Hymenoptera, normal females resulting from fertilized eggs and normal males from unfertilized eggs. However, if eggs are fertilized by sperm carrying the driving B chromosome, they develop into males. The B chromosome causes the condensation and loss of all the chromosomes in the sperm that bears it, with the exception of itself (Nur et al. 1988). This means that the entire paternal A chromosome complement is lost from the zygote, with the result that the zygote is effectively haploid and develops as a male.

The action of this B chromosome makes sense in terms of its own survival. By occurring in males and behaving in such a way that male production is favored, it increases its chances of being passed on. Because males are haploid, all the chromosomes in a male jewel wasp are normally transmitted in sperm. Were the B chromosome present in female wasps, it would only be passed on to half the eggs. The PSR B chromosome is thus a classic example of an ultraselfish genetic element because it manipulates its host's sex ratio to its own benefit and away from the host's optimum.

Conflict between Cytoplasmic and Nuclear Genes

Ultraselfish behavior affecting the sex ratio of bearers is not confined to nuclear genes. Genes inherited in the cytoplasm of gametes, usually just from the female parent, can also cause sex ratio distortion. Indeed, the majority of examples of known sex ratio distorters are cytoplasmic. This is not surprising, given that in most species all cytoplasmic genes, whether they be mitochondrial, chloroplastic, or within inherited symbionts, are derived purely from the parents of one sex, while most nuclear genes are derived equally from both parents. The scope for conflict between cytoplasmic genes and nuclear genes is clear. Nuclear genes, except those on nonpairing parts of the sex chromosomes or on B chromosomes, show no bias to one sex or the other, since both transmit them equally. Cytoplasmic genes, on the other hand, favor the sex through which they are transmitted.

Misbehaving Mitochondria: Cytoplasmic Male Sterility

The elements of this conflict between cytoplasmic and nuclear genes can be illustrated by a phenomenon known as *cytoplasmic male sterility* (CMS) that occurs in some species of hermaphrodite (monoecious) plants. In species showing CMS, some plants are found to be hermaphrodite while others are effectively female, for although they produce both female and male parts of flowers, the male parts fail to produce pollen. They are thus, in reality, male-sterile hermaphrodites. Cytoplasmic male sterility occurs in a wide variety of plant species, including maize, thymes, petunias, plantains, sugar beets, and chives. Early analysis showed CMS to be inherited maternally, suggesting that a cytoplasmic factor might be involved. Using a technique called *restriction fragment length polymorphism* (RFLP) of the mitochondrial and chloroplast genomes, which, in essence, cuts the genetic material up every time a specific sequence of nucleotides occurs, Levings and Pring (1976) were able to show that CMS in maize was always associated with one particular mitochondrial DNA variant.

Mitochondria, and, for that matter, chloroplasts, are thought to be ancient intracellular prokaryotes that were integrated into larger single-celled organisms at around the time of the evolution of eukaryotes. In most species, mitochondria are inherited exclusively from the female parent. Thus, in most hermaphrodite plants, mitochondria are inherited through the ovules but not through pollen. A mitochondrial gene, therefore, has no interest in the production of pollen for its own transmission. Should a mitochondrial gene cause pollen not to be produced, it would spread as long as the lack of pollen production also resulted in an increase in either the quantity or quality of ovules or seeds (pollinated ovules) produced. Increases in the quantity or quality of ovules/seeds may be achieved if resources that would have gone into pollen production are diverted to the ovule/seed production. In addition, as male sterile plants cannot self-pollinate because of the lack of pollen, CMS may enhance seed quality as it enforces cross-pollination.

If there are reasons why CMS may spread, we may ask how mitochondria cause some plants to become male sterile and why other plants remain fully functional hermaphrodites. The mechanism of CMS has been elucidated by studies of petunias. Normally, pollen is produced on the anthers of a flower from a group of pollen mother cells. These mother cells are surrounded by a layer of cells called the *tapetum*. The cells of the tapetum are rich in mitochondria, presumably because the production of pollen cells by division of the pollen mother cells requires considerable energy. The lack of pollen in male sterile petunias results from the deterioration of the mitochondria in the tapetum, so that mature pollen grains cannot develop.

That this cytoplasmic sex ratio distorter is in conflict with the nuclear genome is shown by the fact that in every plant species that has been analyzed in detail, nuclear genes that counteract the male sterile mitochondria and restore pollen production are known. Evidence suggests that a full-blown biochemical war has broken out between mitochondrial and nuclear genes in some species. Specific nuclear male fertility restorer genes counteract mitochondrial types (mitotypes) that cause male sterility. Small changes in the DNA sequences of CMS mitotypes allow escape from the effects of the restorer genes, because the restorer genes only recognize specific mitotypes. Selection then favors the evolution of nuclear genes that recognize the new mitotypes and can trigger the immune system against them. The result is a cycle of an increase in the frequency of a new CMS mitotype, followed by a reduction in its frequency once a restorer with specific recognition of that mitotype arises. Recent studies of CMS in wild populations of the plantain, *Plantago lanceolata*, suggest that there are a great many mitochondrial variants that may induce male sterility and a similar diversity in corresponding restorer genes. The war is never ending. At any one time, mitochondria or nuclear genomes may be in the ascendancy, but neither holds sway for long.

The basic pattern of conflict is similar to that seen with chromosomal meiotic drive. An ultraselfish genetic element, the CMS mitotype, spreads despite the detriment it causes to its host, just as the driving sex chromosome spreads even though it has adverse effects on its carrier. Similarly, just as the harmful effects of drive induce selection in favor of drive suppressors, so the negative effects of CMS on the nuclear genome leads to the evolution of nuclear genes that quash the expression of male sterility mitotypes. There is one crucial difference between the two systems: while there appears to be great diversity in CMS-inducing mitotypes and their corresponding restorers, this does not appear to be the case with drive. In *D. melanogaster* the autosomal segregation distorter *Sd* and its corresponding responder *Res* gene do not show great specificity. Rather, here the effectiveness of the responder gene in suppressing drive appears to be a quantitative trait. The *Res* locus has a region of 120 nucleotide base pairs that may be repeated in exact sequence a number of times. The greater the number of repeats in a *Res* allele, the greater the susceptibility of the chromosome to drive. Thus, chromosomes with a fully insensitive *Res* allele are transmitted at normal Mendelian expectation; that is, with a probability of a half. As the number of repeats of the 120bp sequence increases, the transmission probability of the chromosome declines toward zero.

Not all examples of conflict between the cytoplasmic and nuclear genomes involve genes that can be considered a normal and integral part of the organism's own genome. Some genetic material is passed from generation to generation in the form of intracellular parasites that are inherited in the cytoplasm of female gametes. Just as mitochondrial genes may bias the sex ratio in favor of the females that can transmit them, so inherited parasites may bias their hosts' sex ratios toward females for the parasites' own transmission benefit. The first phenomenon shown to involve such an intracellular parasite is called *cytoplasmic incompatibility*. Although this does not usually cause a bias in the sex ratio, except in haplo-diploid organisms, its description is useful as an example of the behavior of intracellular parasites.

Cytoplasmic Incompatibility

Cytoplasmic incompatibility (CI) was first detected by Ghelelovitch (1952), when crosses between mosquitoes from European, Asian, and American populations failed to produce offspring, or produced offspring only when the crosses were carried out in one direction. For example, crosses of *Culex pipiens* from Hamburg and Ogglehausen populations were normal when females came from Hamburg, but failed when females were from Ogglehausen. The female progeny from Hamburg female × Ogglehausen male crosses, when back-crossed to males from either parental population, were normal, but male progeny only yielded offspring when back-crossed to Hamburg females: with Ogglehausen females, the crosses failed. This observed pattern of reproductive successes and failures demonstrated maternal inheritance, indicating the possibility of the involvement of some cytoplasmic factor; hence the name given to the phenomenon. Eventually, a bacterium of the genus *Wolbachia* was found to be responsible (Yen and Barr 1973, 1974). Since the early detailed work on CI by Laven in the 1950s (see Laven 1967 for review), CI has been reported among many orders of insect, including true bugs (Hemiptera), wasps (Hymenoptera), beetles (Coleoptera), butterflies and moths (Lepidoptera), and true flies (Diptera).

Cytoplasmic incompatibility occurs when a male carrying a CI *Wolbachia* mates with a female that does not bear the same *Wolbachia*. The bacterium is thought to secrete a chemical into the sperm of its host that kills zygotes formed within the female parent if they do not bear the *Wolbachia*. In essence, then, an uninfected female that mates with an infected male is rendered sterile thereafter, even if the female subsequently mates with uninfected males. Other mating combinations are not affected (Table 4.1).

Two patterns of reproductive failure due to CI have been observed. First, as with the mosquitoes described above, some interpopulation crosses are normal if males are drawn from one population, but fail if males are drawn from the other. Second, some interpopulation crosses give rise to few if any offspring, irrespective of which way around crosses are performed. These two patterns suggest that some species of insect harbor one type of *Wolbachia* while others harbor two or more strains. Crosses between populations involving a single *Wolbachia* strain only fail in one direction. However, when the different populations harbor different CI *Wolbachias*, crosses fail in both directions. In *Drosophila simulans*, for example, both unidirectional and bidirectional incompatibility have been recorded, with five naturally occurring incompatibility types having been recorded (Table 4.2).

That a species may harbor more than one strain of CI *Wolbachia* raises the possibility of double infection within an individual. Analysis of populations of the mosquito *Aedes albopictus*, from Houston and Mauritius, has provided evidence that this does occur. The results of exhaustive breeding experiments and

TABLE 4.1
The effect of infection of a single CI *Wolbachia* infection on various mating combinations.

	Infected Female	*Uninfected Female*
Infected male	Normal	Sterile
Uninfected male	Normal	Normal

TABLE 4.2
Incompatibility types and associated infections in *Drosophila simulans*. (uni = unidirectional incompatibility; bi = bidirectional incompatibility.) (After Hoffmann and Turelli 1997.)

Incompatibility Type Notation	*Infection*	*Type of Incompatibility Produced*						*Occurrence*
		R	*S*	*H*	*N*	*A*	*W*	
R	*w*Ri	—						Natural
S	*w*Ha, *w*No	bi	—					Natural
H	*w*Ha	bi	uni	—				Natural
N	*w*No	bi	uni	bi	—			Induced by segregation
A/M	Unnamed/*w*Ma	uni	uni	uni	?	—		Natural
W	uninfected	uni	uni	uni	uni	none	—	Natural
Unnamed	*w*Ri, *w*Ha	uni	?	uni	?	?	uni	Produced by microinjection
Unnamed	*Wolbachia* from *Aedes*	bi	?	bi	?	?	uni	Produced by microinjection

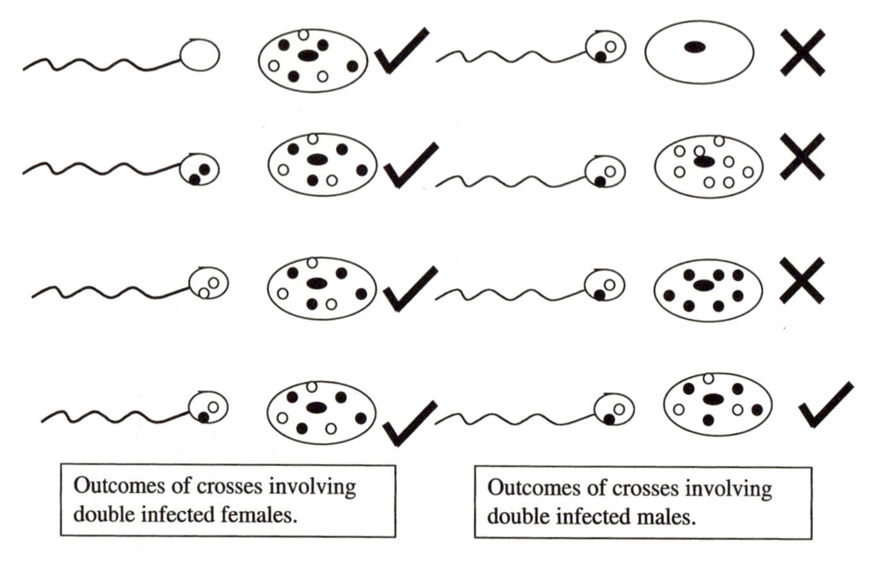

Figure 4.3 Infection by zero, one, or two strains of CI-inducing incompatibility will produce a variety of incompatibility patterns. The diagram shows the outcomes of crosses in which females (left) or males (right) are infected with two strains. (After Siskins et al. 1997.)

molecular genetic analysis suggest that the Houston population harbors two strains of *Wolbachia*, while that from Mauritius harbors only one (Siskins et al. 1995). In the crosses, the pattern of incompatibility was consistent with the Mauritius females being uninfected, even though molecular probes that would specifically reveal the presence of *Wolbachia* showed that a *Wolbachia* was present in this population of flies. The confirmation that two or more strains of CI *Wolbachia* can coexist in the same host individual is interesting because it helps to explain some of the complex patterns of incompatibility previously recorded in mosquitoes (e.g., Magnin et al. 1987) (see Figure 4.3).

Taxonomic Diversity of Species Showing CI

Cytoplasmic incompatibility has been reported as a result of interpopulation crosses of a diverse range of insect species. In *Culex* and *Aedes* mosquitoes, both unidirectional and bidirectional CI have been observed. The same is true of *D. simulans* and several other fruit fly species. In the flour beetle, *Tribolium confusum*, and the alfalfa weevil, *Hypera postica*, unidirectional CI has been detected. Unidirectional incompatibility has also been observed in the plant hopper *Laodelphax striatellus*, and in two species of Lepidoptera, the almond moth, *Ephestia cautella*, and the pierid butterfly, *Eurema hecabe* (Hiroki pers. comm.). It has also recently been detected in the butterfly *Acraea acerata*. Finally, both unidirectional and bidirectional *Wolbachia* CI has been reported in the jewel wasp, *N. vitripennis*, and double infections have also been observed

(Breeuwer et al. 1992). This last case is of particular interest, because as a result of the mechanics of incompatibility, and the chromosomal complements of males and females in haplo-diploid species, here CI *Wolbachia* cause a distortion in the sex ratio (p. 103).

In many of the species in which CI has been observed, researchers have shown that the CI phenotype can be cured by treatment with the antibiotic tetracycline. In addition, in some species, cures have also been achieved by exposing infected strains to high temperatures (around 37°C) for several days.

Despite the diverse range of insect species showing *Wolbachia*-induced CI, the number of species exhibiting the phenomenon is not huge. This is undoubtedly purely the result of lack of work. The majority of species in which CI has been demonstrated are species that either have economic importance as disease carriers or crop pests (e.g., malaria-carrying mosquitoes or pests of grain products in the case of *Tribolium*), or they are well-researched scientific organisms (e.g., *Drosophila* and *Nasonia*). Given estimates from molecular surveys that suggest that *Wolbachia* are present in 10–20% of insect species (Werren et al. 1995a; West et al. 1998; Werren and Windsor 2000), there is a high probability that many additional examples of CI-inducing *Wolbachia* will be revealed in the future. Indeed, examples in both crustaceans and spider mites have recently been reported (Bouchon et al. 2000; Vala et al. 2000), and I expect CI to be reported from many other invertebrate phyla in the near future.

Mechanisms of Incompatibility

The precise mechanics of incompatibility are not well understood. This is partly because the mechanism of incompatibility seems to vary between host species, so that findings with respect to one species should not be assumed to be applicable to others. In *C. pipiens*, incompatibility appears to result from failure of sperm bearing a CI *Wolbachia* to fuse properly with female gametes that lack the same *Wolbachia*. In *Drosophila* and *Ephestia*, embryo development appears to be suppressed at an early stage. Here it has been proposed that *Wolbachia* in eggs of infected females produce a substance that renders them immune to the chemical produced by sperm from infected males (Snook et al. 2000).

The situation in *N. vitripennis* is a little more definite. In this wasp, *Wolbachia* interferes with condensation of the paternal chromosome set during the first mitotic cell division of the embryo. The way that CI *Wolbachia* interacts with the chromosomes of *Nasonia*, causing condensation failure is not known. However, the result is that at this very early stage in zygote development, the paternal chromosome set is lost. All progeny that get the *Wolbachia* thus carry just the maternal chromosome set. They are thus haploid and so all develop as males. This is generally given as a case of a distortion in the primary sex ratio (see p. 47). However, the distortion actually occurs immediately after fertilization and zygote formation, so, strictly speaking, it biases the secondary sex ratio by masculinization of genetically diploid and so initially female zygotes, although this is perhaps splitting hairs a little too finely.

In many species, only young sperm bear *Wolbachia*; mature sperm do not. This is the basis for the argument that the *Wolbachia* both secrete and deposit some chemical into sperm, or possibly remove or modify some feature of sperm during spermatogenesis. However, which one of these arguments, if either, is correct, awaits verification.

Although the underlying mechanics of CI are poorly understood, it is known that *Wolbachia* density influences the degree of incompatibility. Increased *Wolbachia* density is correlated with higher levels of incompatibility. However, comparison between species also shows that the bacterial density needed to cause a specific level of incompatibility varies between host species.

Vertical Transmission Efficiency of CI Wolbachia

The vertical transmission efficiency of cytoplasmic parasites—that is, the proportion of host progeny that inherit the symbiont from their mother—varies between both symbionts and hosts. In the case of CI *Wolbachia*, vertical transmission efficiencies estimated to date are generally high (between 95% and 100%). Evidence of incomplete vertical transmission—that is, that some progeny are completely free of the *Wolbachia*—comes from both mosquito and fruit fly examples. In species in which the vertical transmission efficiencies have been measured both in captivity and in the field, the efficiency of transmission tends to be lower in the field. The existence of naturally occurring antibiotics, climatic factors, particularly temperature, host diapause, and variations in bacterial density levels, may all influence the proportion of offspring of an infected female that inherit the infection. The lower rate of vertical transmission in naturally occurring populations is thus not surprising, since all these factors are liable to be more variable in the wild than in the more controlled conditions in a laboratory.

Effects of CI Wolbachia *Infection on Host Fitness*

It is often assumed that simply bearing a parasitic organism is likely to impose costs on a host in the form of energy needed to sustain the parasite. However, there are many cases in which inherited microorganisms actually enhance the fitness of their hosts. The rationale behind such cases is easy to follow. If the survival and transmission of a symbiont depends on the survival of its host, it is not in the interests of the symbiont to damage its host. Indeed, selection should favor symbiont strains that are beneficial to their host. It is following this logic that we arrive at the idea that microorganisms that are only transmitted vertically should evolve into beneficial mutualistic associations with their hosts. There are many examples of such mutualisms.

Aphids, for example, feed on the phloem sap of plants. Within specialized cells in the body cavity of aphids reside bacteria of the genus *Buchnera*. These are essential to the well-being of their hosts. If the *Buchnera* are killed by antibiotic treatment, the aphids do not grow properly and fail to reproduce. A similar case is seen with the mutualistic gut bacteria (Blattabacteria) of cockroaches and termites. Other examples include photosynthesizing algae living in hydra, nitrogen-fixing bacteria that live in nodules in the roots of leguminous plants, and luminescent bacteria in some fish and mollusks. This being the case, the question of whether ultraselfish cytoplasmic microorganisms impose a direct cost on the individuals that bear them urgently needs attention. The rather sparse current evidence indicates that, in some species, such as *D. simulans*, there is a fitness cost associated with CI *Wolbachia*. In captivity, infected females of *D. simulans* were less fecund than uninfected females (e.g., Hoffmann et al. 1990), although field experiments did not show this effect. Negative effects on host fitness have not been shown in other *Drosophila* species despite being sought, but a reduction in productivity has been recorded for one strain of flour beetle, *T. confusum*, infected with *Wolbachia*.

There is one piece of evidence that suggests that bearing *Wolbachia* may also increase host fitness. *Wolbachia*-bearing males of *T. confusum* appear to gain an advantage over uninfected males through sperm competition. In laboratory tests, Wade and Chang (1995) showed that sperm from infected males had a competitive advantage over sperm from uninfected males. Such an effect would alter the dynamics of the spread of CI *Wolbachia* through a population, the rate of spread being enhanced in promiscuous species.

Field Prevalence Levels

Prevalence levels of CI-inducing *Wolbachia*, that is, the proportion of individuals in a population that are infected, has only been extensively assayed in *D. simulans* and *D. melanogaster*. Population assays on *D. simulans* in California have yielded some extremely interesting data on the spread of CI *Wolbachia* in the wild. Until 1988, infected populations were only found south of the Tehachapi range to the south of the Central Valley. Between 1988 and 1994, there was an extremely rapid increase in the prevalence of infection in the Central Valley. Furthermore, a model, based on the levels of incompatibility and vertical transmission efficiency and assuming that bearing the symbiont imposes no direct cost or benefit on its bearer, produced theoretical predictions close to the observed rates of increase.

Other prevalence data in *D. simulans* and *D. melanogaster* show less perfect agreement with theoretical predictions. This is possibly because it is not always apparent whether CI has reached an equilibrium level in a particular population. This difficulty is exacerbated by data from *D. simulans* suggesting that the precise equilibrium frequency of infection may change over time due to changes in incompatibility levels and vertical transmission rates. Inconsistencies between values observed in the field and those predicted by theoretical models could result from the transportation of fruit flies over long distances when fruit is moved by people. Certainly the northward rate of spread of CI type R, described by Turelli and Hoffmann (1991), which amounted to a rate of migration of around 50km per generation, is too great to be accounted for by the flight of these small flies (see Hoffmann and Turelli 1997 for review).

The CI *Wolbachia* type R has spread rapidly around the world, and increased in prevalence very quickly in many countries. The agent of this type of CI is notated as *w*Ri. It now occurs at high prevalence throughout North America, except in southern Florida. High prevalences of *w*Ri have also been reported in Costa Rica, Uruguay, Zimbabwe, Portugal, and Spain. Lower frequencies have been reported in Israel, France, and Italy in the last decade, and it would be helpful to remonitor these areas to discover whether measured levels were transitional as the *Wolbachia* spread through to equilibrium or fixation.

The Taxonomic Relationships of CI Wolbachia *and Their Hosts*

All the agents known to cause CI that have been analyzed and identified using molecular genetic DNA sequence technology belong to the *Wolbachia pipientis* complex. This complex represents a widespread group of α-Proteobacteria. Members of this complex are known to infect hosts from all the major orders of insects as well as some other arthropods (mites, spiders, isopods) and phyla (nematodes). Well over 200 invertebrate host species have been identified and new hosts are being discovered on a regular basis.

The *Wolbachias* are a truly remarkable group of bacteria in their interactions with their hosts. Apart from causing CI and incidentally masculinization in *N. vitripennis*, members of this complex are known that cause feminization (chapter 5), male-killing (chapter 6), and parthenogenesis (chapter 8). In addition, some appear to have no overt effect on their hosts, while some are beneficial. Phylogenetic analysis has shown that, within arthropod hosts, the complex involves two main lineages that split about 50 million years ago (Werren et al. 1995b). This type of analysis has also shown that the evolutionary histories of *Wolbachias* and their hosts are different. In other words, *Wolbachias* in very taxonomically different hosts may be more similar to one another than *Wolbachias* in very closely related hosts. This suggests that interspecific horizontal transmission of *Wolbachias* must occur at times. That said, virtually nothing is known about the patterns and frequencies of horizontal transfer of these bacteria in the wild, except that there is some evidence that *Wolbachia* can transfer horizontally between parasitic insects and their hosts (p. 193).

The lack of phylogenetic correlation between *Wolbachias* and their hosts extends to their effects on their hosts. *Wolbachias* that use different strategies to manipulate host reproduction do not group together taxonomically. For example, the CI *Wolbachias* do not form a single taxonomic group compared to those that induce parthenogenesis. On a smaller scale, there is evidence to suggest that the male-killing *Wolbachias* of the two-spot ladybird, *Adalia bipunctata*, are most closely related to the CI agent of the flour beetle *Tribolium confusum*. Conversely, the male-killer of *Acraea encedon* aligns with the parthenogenesis-inducing *Wolbachia* of the egg parasitoid wasp, *Trichogramma deion*. One explanation for this lack of tight grouping of *Wolbachias* with particular effects on hosts is that the various strategies employed by *Wolbachia* have each evolved independently more than once. This is perhaps the most likely explanation, but others exist as well. Different strategies may arise due to different selection imposed by a host's behavior and ecology. Particulars of a host's genome may also provide favorable openings or impose selective constraints on *Wolbachia*, so that the host's genome, or some part of it, acts as a critical determinant of the manipulation expressed. It is also feasible that, within a host species, the evolution of a suppressor system against one type of manipulation leads to the evolution of an alternative manipulation that is not suppressed. Recent work using micro-injection techniques to transfer *Wolbachia* between hosts has shown that bacteria that cause feminization in one host species may cause a different phenotype, such as male-killing, in another host (see p. 123).

Cytoplasmic Incompatibility and the Sex Ratio

Most CI-inducing *Wolbachia* do not affect the sex ratio of their hosts. However, in the case of haplo-diploid hosts, there is a strong influence on sex ratios, with males being favored. Saul (1961), who reported CI in *N. vitripennis*, first discovered this. In this parasitoid of fly puparia, local mate competition causes a natural female bias in the sex ratio produced by females (see p. 44). In inbred laboratory lines, Saul found that females that mated produced daughters about 85% of the time. Unfertilized females produced only haploid males. When the two strains were crossed in one direction, the progenic sex ratio was normal (85% female), but in the other direction a severe bias in favor of males resulted, suggesting CI. Saul showed that the trait was maternally inherited. Later, genetic analysis showed that in male-biased progeny of mated females, the CI factor caused the loss of the complete paternal set of chromosomes, so that haploid and thus male zygotes resulted (Ryan et al. 1985).

The CI factor was initially ascribed to a bacterium when it was discovered that the trait could be cured by application of antibiotics to host larvae or puparia by micro-injection (Richardson et al. 1987). Molecular genetic analysis of the 16S rDNA gene, which is often used to determine bacterial identity, revealed the cause of the incompatibility to be a *Wolbachia*. Indeed, Breeuwer and his coworkers, finding a variety of 16S sequences in some *Nasonia* lineages, gave as one plausible explanation the coexistence of several *Wolbachias* within the same host individual (Breeuwer et al. 1992). This was confirmed when Perrot-Minnot discovered two distinct *Wolbachias* in some *Nasonia* strains. Other strains carry only a single infection. Thus, both bidirectional and unidirectional incompatibility should be observed in this wasp, an expectation that was confirmed by Perrot-Minnot and his colleagues (Perrot-Minnot et al. 1996).

It is interesting that in the case of both B-chromosome PSR and *Wolbachia*-induced parthenogenesis, the mechanism for the production of an excess of males depends on the loss of the paternal chromosome set. It may be a long shot, but it is interesting to speculate that the gene(s) involved in each of these cases might be the same. Currently, there is a wide-ranging debate over the extent to which genetic material can move around the genome and between genomes. The extent of recombination between mitochondria and between different strains of *Wolbachia* have both been the subject of discussion with empirical evidence that some recombination occurs in both recently being obtained (Jiggins et al. 2001a; Werren and Bartos 2001). In addition, the presence of identical or almost identical DNA sequences in mitochondria and nuclear chromosomes suggests occasional genetic exchange between the cytoplasmic and nuclear components of the genome. It is thus feasible that a gene or genes causing CI in *Nasonia* have become transferred to a nuclear supernumerary chromosome, thereby favoring its transmission. The precise gene(s) responsible for CI in *Wolbachia* are not known and may be different in different hosts. However, it would be of great interest to discover the gene or genes responsible for both CI and PSR in *Nasonia* and to compare their exact sequences and modes of action.

Cytoplasmic Influence on Fertilization in Haplo-Diploids

In this chapter, we have already seen two ways (B-chromosome PSR and *Wolbachia*-induced CI) in which ultraselfish genetic elements distort the sex ratio of the jewel wasp *N. vitripennis*. There is a third way. This is called *maternal sex ratio* (*msr*) and is perhaps the least understood of the sex ratio–distorting phenomena of this wasp. *Msr* is a maternally inherited cytoplasmic factor that results in an increase in the rate of fertilization of eggs and so increases female production (Beukeboom and Werren 1992). In some instances as many as 97% of progeny of *msr* females are daughters. The cytoplasmic factor that causes *msr* is not known. Searches for bacteria associated with the trait have been fruitless and it is usually conjectured that a virus or a specific mitochondrial mutant is responsible. It is interesting that different factors, one cytoplasmic (*msr*) and one nuclear (PSR), pull the host's sex ratio away from the optimum in opposite directions. Here is a very obvious example of warfare within the genome.

Cytoplasmic Sex Ratio Distortion in ZZ/ZW Insects

One final type of cytoplasmic sex ratio distortion seems plausible. In species with ZZ/ZW sex chromosomes, irrespective of whether femaleness is determined by the W chromosome or the lack of two Z chromosomes, a maternally inherited cytoplasmic agent would increase its chances of transmission if it caused W chromosomes to enter more than half the female gametes. Should such a cytoplasmic agent have reasonably high vertical transmission, it would have the same pattern of inheritance as a driving W chromosome (or the lack of a sex chromosome in species with ZO females). Distinguishing a cytoplasmic agent that favored transmission of W chromosomes into female gametes from a W chromosome driver would thus be almost impossible using normal genetic cross-analysis.

This type of cytoplasmic agent could be responsible for data produced in the early part of the 20th century by Doncaster (1913, 1914), working with the magpie moth, *Abraxas grossulariata* (Figure 4.4a). Doncaster's observations began with the discovery of a female moth that produced only daughters. In subsequent crosses, from the strain generated from this female, about half the daughters themselves produced severely female-biased progeny. The rest produced normal sex ratios. Examination of chromosome complements showed that females that produced all or predominantly daughters had 55 chromosomes, while normal females had 56. This suggests that the abnormal females may have been ZO while normal females were ZW. Examination of meiosis in females revealed that in females with 55 chromosomes, most female gametes had 27 chromosomes, with one chromosome being left on the metaphase plate. Analysis of the details of Doncaster's crosses suggests that neither parthenogenesis nor male-killing can account for his results. In the case of parthenogenesis, some crosses involved males carrying color pattern genes and these were inherited. In the case of the death of male embryos, the number of female progeny in female-biased families was double that born to normal females, suggesting that a sex-related bias in mortality was not involved (Hurst 1993).

We are left with two possible causes of the rarity of males in this strain: either sex chromosome meiotic drive or a cytoplasmic factor favoring the lack of migration of the sex chromosomes during meiosis. The evidence provided by Doncaster does not allow these two possibilities to be distinguished. However, given the absence of any confirmed case of sex chromosome meiotic drive in the Lepidoptera and the argument that such drive in female heterogametic species is likely to be detrimental (p. 89), this may be a case of a cytoplasmic agent preventing Z chromosomes from entering eggs.

Similar female biases have been reported in a small number of other Lepidoptera. In particular, all-female broods have been observed in both the meadow brown butterfly, *Maniola jurtina* (Figure 4.4b) (Scali and Masetti 1973), and the drinker moth, *Philudoria potatoria* (Figure 4.4c) (Majerus 1981), and in neither case does male-killing seem a tenable explanation of the data published. Both are potential examples of cytoplasmically induced meiotic drive, and reinvestigation would be timely.

Why Are Genetic Law Breakers the Exception Rather Than the Rule?

In this chapter, we have seen that the genetic makeup of higher organisms is rife with the potential for genetic conflict. Sex chromosomes fight, autosomes

a

b

c

Figure 4.4 Unexplained female-biased progenic sex ratios have been recorded in a number of Lepidoptera. a) The magpie moth, *Abraxas grossulariata*; b) the meadow brown butterfly, *Maniola jurtina*; c) the drinker moth, *Philudoria potatoria*.

may become spiteful toward their homologues to increase their own chances of transmission, B chromosomes cause the destruction of other chromosomes inherited with them, and both mitochondria and a variety of microorganisms that live in the cytoplasm of their hosts' cells manipulate the reproduction of their hosts shamelessly to their own ends. Strategies employed include emasculation, feminization, masculinization, assassination of nonbearers, killing of male hosts, or getting rid of sex altogether. The "fair" part of the genome, that is, that part that is inherited equally from both parents, tries to fight back through responders, suppressors, rescue genes, and the like.

Despite the range of ultraselfish behaviors of genetic elements described in this chapter, and I have no doubt that other ways in which genes or genetic elements that abuse their hosts to their own benefit await discovery, most genes do behave themselves. Despite all the potential for anarchy that exists in the genomes of most organisms, most genes exist and are inherited in apparent harmony. Most nuclear genes are inherited in accordance with Mendel's laws, and only a few cytoplasmic genes are known to misbehave. On current evidence it seems that chloroplasts do not favor the sex of the parent through which they are inherited. Mitochondria only favor the sex through which they are inherited in a minority of plant species. Female-biased sex ratios are not, as far as is known, caused by mitochondria in animals, despite the fact that the transmission rate of mitochondria would be enhanced if they were. We know of relatively few examples of sex chromosome meiotic drive or B chromosome drive. So we need to ask why, in spite of all the possibilities for nastiness within the genome, do most genes play fair?

There are two likely answers to this question. The first derives from the conflict between different elements of a host's genome. An ultraselfish strategy that is in the interests of one small part of a genome is unlikely to be in the interests of the rest of the genome. As "the rest" will be much larger, there should be considerable scope for selection to favor some genetic mechanism that brought the lawless element of the genome back into line. The existence of suppressor systems that act against most of the law breakers described in this chapter, and the ultraselfish inherited parasites mentioned in forthcoming chapters, supports this idea of law and order in the genome.

The second answer is simple. If law-breaking genes arise and spread in an organism, so that they are detrimental to the organism, then, if mechanisms do not arise to ameliorate or suppress their detrimental effects, they may just drive their hosts, and themselves, to extinction. As Darwin stressed, the environments inhabited by all organisms are harsh. If the cost imposed by law-breaking genes on their bearers is too great, the gene carriers will simply be outcompeted in this incredibly competitive biotic world. What are left are those genetic elements that manage to get on with each other.

There are two other rather unsatisfactory and less likely possibilities with respect to the question of why most genes play fair. The first can be couched as a question: what would happen if an individual were host to more than one type of outlaw? It is unlikely that we would find the genetic equivalent of "honor among thieves." In fact, we have seen in the case of *N. vitripennis* that several outlaws have invaded the same host. These outlaws should be antagonistic toward one another. It appears that, just as with mob leaders, "this town (or host) ain't big enough for the both of us." Other cases in which more than one outlaw have been recorded include the woodlouse *Armadillium vulgare* (p. 125), the two-spot ladybird *A. bipunctata* (p. 139), and the butterfly *A. encedon* (p. 159). Detailed investigations of how the different genetic outlaws interact in these species would be of great value.

The other possibility is that we may simply not yet know the full extent of genetic lawlessness. The range of known genetic outlaws and our knowledge of them has increased greatly over the last two or three decades. Some, particularly those that are difficult to detect, are undoubtedly going to be much more common than we currently suppose. In this context, we may consider *Wolbachia* just one group of many intracellular inherited bacteria. Interest in this group, because of its ultraselfish behavior toward its host, has increased dramatically over the last decade. From a molecular search of neotropical insects using a *Wolbachia*-specific genetic probe, John Werren and his colleagues came to the astounding estimate that between 10% and 20% of all insect species harbor *Wolbachia* (Werren et al. 1995a). Other similar analyses have confirmed this result (West et al. 1998; Werren and Windsor 2000). Recently, Jeyapraskash and Hoy (2000), using a method of molecular genetic analysis called long PCR amplification, have suggested that this figure may be an underestimate. In a survey they found *Wolbachia* in 76% of 63 arthropod species assayed, although the methodology has received some criticism. Because ultraselfish bacteria of at least six other major groups (p. 137), in addition to some protists (pp. 119 and 134), are also known, the proportion of invertebrate species that are host to intracellular parasites is probably a majority.

If this is the case, then together with meiotic drivers, B chromosomes, and other elements that do not play the Mendelian game, the majority of organisms are likely to have some renegade genes. As these can have profound effects on biological features that are fundamental to the understanding of any organism's ecology, behavior, and evolution, it is imperative that efforts be made to identify these genetic renegades and understand their stratagems and effects.

The Realization of the "Pretty Improbable"

Discussing genetic outlaws about twenty years ago, Richard Dawkins (1982) played a thought experiment that he admitted at the time was "pretty improbable." He wrote:

> Imagine a gene on a Y chromosome which makes its possessor kill his daughters and feed them to his sons. This is clearly a behavioural version of a driving Y chromosome effect. If it arose it would tend to spread for the same reason, and it would be an outlaw in the same sense that its phenotypic effect would be detrimental to the rest of the male's genes. (p. 143)

TABLE 4.3
The manipulative strategies of known ultraselfish inherited symbionts.

Manipulative Strategy	*Causative Agent(s)*
Cytoplasmic incompatibility	*Wolbachia*
Feminization	*Wolbachia*, Protists
Parthenogenesis induction	*Wolbachia*
Male-killing	Microsporidia, α-proteobacteria, γ-proteobacteria, Flavobacteria, Mycoplasmas

We can perhaps extend this unlikely scenario by reversing the sexes. So a son-killer gene on a W chromosome in a female heterogametic species would spread in the same way. Furthermore, the same is true of a gene in the cytoplasm that is inherited through one sex, say, females, only. But now we have moved from a theoretically "pretty improbable" mind game to fact, for many son-killing genetic elements are known. These are the male-killing microorganisms that have been recorded in many insect species and are likely to be discovered in many other groups of invertebrates in the future.

Male-killing is just one of the types of reproductive manipulation that has been adopted by inherited cellular symbionts (Table 4.3). We have already discussed one of the other strategies, CI. This only affects host sex ratios in the limited instances where it occurs in species in which ploidy levels determine sex. Other manipulations involve feminization and induction of secondary asexual reproduction. These, together with male-killing, cause severe distortions in their host sex ratios, placing them in direct conflict with the nuclear genomes of their hosts. It is the extraordinary microbes that adopt these strategies and the effects that they have on the evolutionary biology of their hosts that I will focus on in Part II of this book.

PART II

Ultraselfish Symbionts

CHAPTER 5

Feminization

> And the Lord God caused a deep sleep to fall upon Adam, and he slept:
> And He took one of his ribs, and closed up the flesh instead thereof;
> And the rib, which the Lord God had taken from man made He a woman.
> —The Book of Genesis, Chapter 2, verses 21–22

Summary

Sex changes are known in a number of organisms. In some, sex change is part of the normal life history and is programmed within the normal genome of a species. In others, however, the sex determination system of a species is usurped by symbiontic microorganisms. In such cases, genetically male hosts are changed into functional females. This feminization is to the benefit of the causative symbiont because the symbiont is maternally inherited. The aim of this chapter is to consider the interactions that occur between these microbial feminizers and their hosts.

Examination of the interactions between microbial feminizers and their hosts suggest that these interactions are highly antagonistic, with symbiont and host evolving mechanisms to manipulate the sex determination system for their own interests. The adaptations and counteradaptations that have evolved represent outstanding examples of the products of an evolutionary arms race.

From Adam to Eve, or Vice Versa

The first alleged human sex change was the production of the first female from male tissue: Eve being created from Adam's rib.

In humans, as in other mammals, the basic difference between males and females is based on the type of gamete produced. Individuals divide easily into those that produce sperm and those that produce eggs. Difficulties only arise in sterile individuals who produce neither, and then the primary sex organs indicate sex. However, if other sexual characteristics are considered, the division between maleness and femaleness may be less sharply defined. Degrees of development of many morphological and behavioral characteristics that are considered either feminine or masculine depend to a considerable extent on the balance of various sex hormones produced. Environment and experience also influence some sexual traits, particularly behavioral traits. Indeed, some individuals whom scientists would define as male on the basis of the gametes that they produce, have at least some female behavioral traits, occasionally to the extent that they feel mentally female but are trapped in a male body. The reverse also occasionally occurs, although less commonly. Modern medical practices, through hormonal treatment and surgery, can help such people who may undergo a so-called sex change operation. Feminization is more common than masculinization. Such operations can be extremely successful in all ways except one. No reproductively functional person of one sex has been transformed into a reproductively functional individual of the opposite sex.

This is not the case in all species. Indeed, we have already seen that in some fish, functional females can transform into functional males if the male in their group dies (p. 79). In other fish, such as the hermaphrodite black hamlet fish, *Hypoplectrus nigricans*, members of a pair take turns playing the male and female roles (Fischer 1980). In this species, there is an element of trust between the fish because of the much greater expenditure of energy in taking the female role. If a fish that has played the easy male role in one reproductive season does not adopt the female role next time around, then its partner will desert it. Of course, the fish that plays the male role the first time could cheat by not reciprocating and producing eggs for its partner to fertilize the next time. However, this rarely happens because the cost of having to find a new mate outweighs the expenditure in producing a batch of eggs. Sex change here, at least from the point of view of which sexual role is played at any particular time, thus seems to be a good example of a tit-for-tat game strategy.

We have already seen (p. 79) that sex changes within an individual's life are the norm in some organisms, with change from female to male being commoner than the reverse in vertebrates. The opposite is the case in invertebrates. The direction of sex change seems to relate to whether males or females are generally larger at maturity, with transitions usually being toward the larger sex. For example, in the slipper limpet, *Crepidula fornicata*, sexual life usually begins as a male. Larval limpets have to settle on a hard substrate and if they settle on a rock, they rather quickly develop as females. However, the majority of larvae settle on other limpets. This is not just by chance. Females send out pheromones to attract them, and a stack of limpets develops. Larvae that settle on the back of other limpets initially develop into males. Examination of one of these stacks shows that the larger, older limpets at the bottom of the pile are females, while the younger, smaller ones on top are males. In the middle of the stack may be a limpet in the process of changing from male to female, for these limpets are sequential hermaphrodites. The males on top mate with the females below by use of a very long intromittent organ that is extended down the stack past any males in between.

This theme, relating the direction of sex changes in different animal taxa to the size of the sexes, can be extended to plants, which for these purposes are more like invertebrates than vertebrates. A remarkable eastern American plant, the jack-in-the pulpit, *Arisaema triphyllum*, illustrates the point. This plant, like the common slipper limpet, is a sequential hermaphrodite, starting life as a male, and transforming into a female later in life. Amazingly, some plants then change back from females into males, but only if circumstances dictate that the size of the plant has been reduced. The flowers of any individual plant each year contain either male pollen-bearing anthers or female seed-bearing ovaries and stigma. It is the size of the plant that determines which. In a superb study, David Policansky (1981) measured the size, seed, and pollen content of over two thousand jacks-in-pulpits. Using this data, he constructed a theoretical model to predict the height that plants should be to make forming female sexual parts advantageous. His prediction from this model was 398mm. Further survey work then showed that the change in the wild occurred at 380mm, which, given the statistical variance in his sample, represents a very close agreement with his prediction. By marking plants and checking them year after year, Policansky was also able to show that some plants changed from female to male if their growth was stunted by being eaten or being shaded by competitors. A transition from female to male also occasionally resulted after a female plant had produced an exceptionally large crop of seeds in one year. This was presumably because "her" resources were so depleted by this excessive bout of reproduction that the following year "he" could not reach the threshold size to obtain a benefit from being female.

Microbe-Induced Sex Change

In the above cases, the transition from one sex to the other is programmed by the organism's own genes. However, in some invertebrates the transformation results not from the expression of their own genes, but from the activities of microbes living within the cytoplasm of their cells. To date, two groups of inherited microorganisms have been identified that manage to change the sex of their host from that predetermined by their hosts' sex chromosomes. In both cases it is individuals that have male sex chromosomes that are turned into fully functional females.

The first of these of microorganisms is *Wolbachia*, which we have already encountered causing CI in a variety of insect species. For some time, it has been known that this bacterium has the ability to feminize woodlouse species. More recently, and in some ways more fascinatingly, Japanese workers have shown it to be implicated in the feminization of two species of moth, the Asian corn borer, *Ostrinia furnacalis*, and the Aduki bean borer, *Ostrinia scapulalis* (Kageyama et al. 1998; Sasaki et al. 2000). The second group of feminizers does not involve bacteria, but primitive protozoa, microsporidians and the closely related paramyxidians, which live within another group of crustaceans. In essence, what both of these groups of microorganisms do is turn genetic males into females.

Amazingly, in some populations this feminization has completely usurped the normal chromosomal sex determination system, with all members of the population having the usual male sex chromosomes, females being genetic males that have the feminizer and so are female, while males are simply uninfected genetic males. In other populations, two types of females exist, genetic females that may or may not harbor the feminizer and genetic males that have been feminized. Thus, some females are homogametic, while others are heterogametic.

Horizontal and Vertical Transmission

It is worth exploring the reasons why these microorganisms manipulate their hosts in this surprising way. For microorganisms that live inside and depend on their hosts for resources, transmission between hosts may be achieved in two different ways. First, transmission may be achieved by simply moving between individuals. Such transmission may be between hosts that come into direct contact, as for example in sexually transmitted diseases, or via an intermediary vector, such as a mosquito vectoring the malaria parasite. Alternatively, they may simply be released into the environment in host feces or in aerosol form through coughing or sneezing, to be picked up by new host individuals, as is the case for the microbes responsible for typhoid and influenza, respectively. Transmission in these cases is said to be horizontal between individuals, irrespective of their age.

The contrasting mode of transmission is "vertical" transmission, when the microbe is transmitted from parent to offspring. As has already been described, this is most usually achieved within the cytoplasm of cells, leading to maternal transmission of the intracellular microbes. This maternal transmission gives the rationale for the feminizing effects of these microbes and for various other methods by which microbes distort sex ratios of their hosts in favor of females.

Options for Maternally Inherited Symbionts in Male Hosts

Sperm of most organisms contribute little if any cytoplasm to offspring. Therefore, a microbial symbiont that gets into a male zygote has a problem because it cannot be transmitted to its offspring through sperm. What options can we then imagine for it?

One strategy that could be adopted by symbionts that cannot be vertically transmitted through males is simply to ensure that they do not get stranded in a male zygote. This is feasible, at least in female homogametic species, if the symbiont in a female gamete can evolve a mechanism to bar fertilization by sperm carrying a Y chromosome. This type of tactic might be particularly appropriate to symbionts whose hosts fertilize externally or just before oviposition, since, for these symbionts, some other strategies are not available. I know of no case in which such a mechanism has yet been found, but given the increasingly sophisticated cryptic female choice behaviors that are now being unmasked (see p. 19), I suspect it is only a matter of time.

An alternative that is available to female heterogametic species is to ensure that all eggs are formed by W-bearing gametes. Meiosis in males and females differs. In males, the products of meiosis are four sperm, each sex chromosome in the original diploid cell being present in two of the products. In females of most species, however, only one egg is produced for each parent cell that undergoes meiosis, the other three haploid products together giving rise to the yolk of the ensuing egg. Here, then, where females are heterogametic there is the possibility of symbionts causing a bias in the sex ratio of offspring by biasing which meiotic product becomes the female gamete in favor of those bearing a W, rather than a Z chromosome. Again no proven case is currently known, but such instances would be difficult to detect, since they would resemble meiotic drive with respect to their phenotypic effect, a female-biased primary sex ratio resulting.

Intriguingly, a female-biased sex ratio in an insect with heterogametic females has recently been found to be associated with a bacterial symbiont. Francis Jiggins (2000) showed that some families of the African butterfly, *Acraea eponina* (Plate 4a), were heavily female biased, with some females only producing daughters. Molecular genetic analysis using two genes (16S rDNA and *wsp*) showed that the female-biased lineages harbored a *Wolbachia*, while lines that produced normal sex ratios did not. By treating *Wolbachia*-bearing lines with antibiotics, the SR trait was "cured." Breeding experiments then showed that the egg hatch rates of the female-biased families did not differ significantly from the lines producing a 1:1 sex ratio, suggesting that the female bias affected the primary rather than the secondary sex ratio. These observations, as far as they go, are exactly what would be expected if the *Wolbachia* were causing W rather than Z-bearing meiotic products to form the eggs. Other explanations, such as late male-killing (p. 134), could explain the data and detailed cytological examination and assessment of mortality levels in infected and uninfected families will be needed to distinguish between alternatives. Irrespective of whether the sex ratio distortion in *A. eponina* is found to be a case of bacterially induced sex chromosome drive, I hope that any further discoveries of primary sex ratio biases that are initially attributed to meiotic drive will be probed to detect whether a bacterial or any other cytoplasmic element is associated with the phenomenon.

Preventive strategies, ensuring that the symbiont does not get stranded in males, are effectively sidestepping the problem. What if the symbiont does get stranded in a male? What, then, are its options? One possibility would be for it to replicate as fast as possible with the hope of killing its host, so that when its host decays, the symbiont will be released into the environment where it may infect another individual. Symbionts within males thus have horizontal transmission as an option. The effectiveness of this strategy will, of course, be limited by the symbiont's ability to survive outside its host and the likelihood that another host will be in the right place at the right time. Despite these limitations, there is one group of microorganisms that are known to have evolved this male-killing strategy. These are a group of microsporidians that infect mosquitoes (see p. 134).

A second fate for a symbiont trapped in a male host is to kill its host in such a way that the death of the host (and the symbiont) is of benefit to the microbe's own kin. This too is a strategy that is adopted by some microbes, in this case a wide variety of bacterial species (see chapter 6).

A third option is to migrate from zygotes destined to be males into zygotes that are genetically female while the zygotes are in close proximity; for example, before they are released by their mother. This is unlikely to be possible in species with external fertilization or in male heterogametic species in which fertilization occurs just before eggs are laid.

The fourth option is to turn genetic males into functional females that have the ability to lay eggs and so transmit the symbiont. As previously described, this feminization of hosts has evolved in at least two types of microorganism. In diploid hosts in which sex is ancestrally determined by sex chromosomes, females thus can be of two types, those that are normal genetic females and those that are feminized males. This means that some females will be heterogametic and some will be homogametic, irrespective of whether males or females are originally the heterogametic sex. Males, on the other hand, will only be of one ancestral type.

Feminizing Microbes

The most fully studied case of symbiont-induced feminization occurs in the woodlouse, *Armadillium vulgare*. The normal sex determination in this species is based on sex chromosomes, with females being heterogametic (WZ) and males homogametic (ZZ) (Juchault and Legrand 1972). Early studies showed that some females produced strongly female-biased progenic sex ratios and that the trait was maternally inherited. Careful observation failed to reveal significantly more mortality among male as compared to female progeny of females exhibiting the trait, suggesting that the bias in the sex ratio did not result from the death of males. Using delicate micro-injection techniques to insert tissues from infected females into normal females that lacked the trait, Legrand and Juchault (1970) were able to show that the trait could be artificially transferred, suggesting that a symbiont might be involved. This suggestion was verified when a bacterium was detected in females showing the trait, but not in males (Martin et al. 1973).

Many bacteria are temperature sensitive and cannot survive at high temperatures. As an aside, this is the main reason that many warm-blooded species elevate their body temperature, in the form of a fever, in response to infection by bacteria. Juchault and his colleagues tested whether the bacteria in the infected woodlice were temperature sensitive by submitting them to a period at 30°C. This worked, the bacteria gradually being destroyed. Interestingly, females submitted to this treatment and freed from bacterial infection subsequently produced strongly male-biased progeny. The simplest way to explain this result is if the females in question were all genetic males (i.e., ZZ); that is, the female-biased sex ratio trait resulted from the feminization of genetic males by the bacterium. The expected result of mating these females with normal males (also ZZ) would be to produce just male offspring, with females only resulting if the cure was not perfect so that a few of the progeny still retained the feminizer. This expectation was realized (Rigaud et al. 1991).

Subsequent examination of the chromosomes of wild populations in which the bacterium had been found revealed the startling fact that all were genetic males, having two Z chromosomes (Juchault et al. 1993). The W chromosome had been deleted completely from these populations. In these populations, therefore, the control of sex determination has changed from sex chromosome–dependent to symbiont-dependent. All individuals are ZZ, with those carrying the bacterium being female and those lacking it being male. The vertical transmission efficiency of the symbiont, which molecular identification showed to be a *Wolbachia*, is imperfect, with approximately 10–20% of progeny being uninfected and so male.

The feminization process produced by *Wolbachia* in woodlice is usually complete, but occasionally intersexual phenotypes do occur. The existence of these has been speculatively explained as being due either to low bacterial density in some individuals or to lack of exact synchrony between the feminizing activity of the bacteria and the timing of sexual tissue determination.

The factors that affect sex determination and sex ratio in *A. vulgare* have been under investigation in France for nearly 40 years. The superb work of J. J. Legrand, Pierre Juchault, Thierry Rigaud, Gilbert Martin, and Jean-Pierre Mocquard has shown, as is so often the case in biology, that once a story is studied and dissected in minute detail, more and more complications are found. In this case, the simple story of sex determination of *A. vulgare* being usurped by a feminizing bacterium for its own benefit is complicated by the existence of a second, nuclear feminizer and a nuclear masculinizing gene: two more genetic outlaws. I will return to these complications presently to examine their effect on population sex ratios in this species.

The Diversity of Feminizers and Their Hosts

Feminizing microbes have now been recorded and identified from a number of species of crustaceans and from two insects. Within the Crustacea, two distinct taxonomic groups of feminizers have been found. In isopods, including the above-mentioned woodlouse, feminization is caused by *Wolbachia* (Rousset et al. 1992). *Wolbachia* also causes feminization in another group of crustaceans, the Amphipoda, but here most feminization is the product of primitive eukaryotes.

On current evidence, the taxonomic split between the two systems in the Crustacea seems almost complete. In a survey of 80 species of crustaceans, 22 species were found to harbor *Wolbachia*. All of these were isopods (Bouchon et al. 1998) with the exception of two, in the amphipods *Orchestia gammarellus* and *Talorchestia deshayesii*. These are intertidal amphipods and live in close proximity with coastal isopods. It is considered probable that horizontal transmission of feminizing *Wolbachia* between crustaceans of these two classes has occurred where these live under the same ecological conditions.

In most of the *Wolbachia*-bearing isopods in which the effect of the *Wol-*

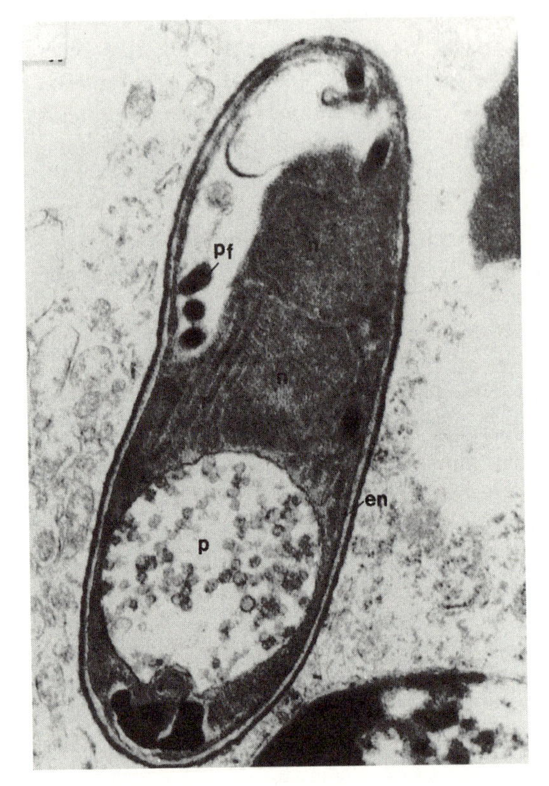

Figure 5.1 Transmission electron micrograph of a microsporidian *Nosema granulosis* spore in a follicle cell of *Gammarus duebeni*. (en = thin endospore; n = diplokaryotic nucleus; r = rows of polyribosomes; Pf = short 3 coiled polar filament; p = granular polarplast.) (Courtesy of Alison Dunn.)

bachia on its host has been ascertained, it causes feminization, although in one, *Porcellio dilatatus*, it induces CI (Bourtzis and O'Neill 1998).

Nonbacterial feminizers are confined to the Amphipoda, the brine shrimp, *Gammarus duebeni*, being infected with two Microsporidia, *Nosema* sp. (Figure 5.1) and *Octosporea effeminans* (Smith and Dunn 1991), and *Orchestia gammarellus* being host to the paramyxidian, *Paramartelia orchestia*, which is closely related to microsporidians (Ginsburger-Vogel and Desportes 1979).

The rough distinction between the two systems appears quite strong, for artificial transfer experiments using micro-injection are successful within species and to some extent within the isopod and amphipod classes, but not between classes. As an illustration, micro-injection of the feminizers of *A. vulgare* (*Wolbachia*) and *G. duebeni* (Microsporidia) into uninfected recipients of each other failed (Dunn and Rigaud 1998). Some of the results from artificial transfer experiments hint that a particular host and its feminizer may have coevolved. The feminizing *Wolbachia* from *Chaetophiloscia elongata*, which has been shown to have very similar DNA to the feminizing *Wolbachia* of *A. vulgare*, when successfully transferred into *A. vulgare*, failed to produce a feminizing effect (Bouchon et al. 1998). This suggests that the mechanism of feminization may depend as much on the host as on the bacterium. This would make intuitive sense if the process of feminization relied on exact DNA sequences of particular genes specific to particular hosts (Dunn and Rigaud 1998).

The different feminizers vary in their distribution within their hosts and in how common they are in their hosts. Taking distribution within hosts first, *Wolbachia* and paramyxidians tend to be scattered widely throughout the tissues of their hosts (Martin et al. 1973; Ginsburger-Vogel and Desportes 1979), while Microsporidia are present exclusively in the female germ line. From an elegant series of observations, Melanie Hatcher, Alison Dunn, and C.M.N. Tofts came to the conclusion that in *G. duebeni*, Microsporidia may have the ability to identify, and therefore preferentially infect, early stage germ line tissues (Hatcher et al. 1997). While tending to this thesis, they also note that, if Microsporidia in germ line cells replicate much faster than those in somatic cells do, the observed distribution pattern could be explained. Whichever is the case, they argue that the germ line–specific microsporidian distribution in *G. duebeni* is probably a result of the relatively slow replication rate of the Microsporidia, making specialization to transmitting cell lines essential to maintain reasonably high vertical transmission efficiency.

The prevalences of the various crustacean feminizers vary widely. In the brine shrimp *G. duebeni*, feminizing Microsporidia have been found in all populations surveyed, with prevalences ranging from 5% to 30%. By contrast, most populations of *A. vulgare* lack feminizing *Wolbachia*, although here the presence of another nonmicrobial feminizer (the *f* factor, see p. 126) complicates the situation. The reasons for the variability in feminizer prevalences, and particularly the low prevalence of feminizing *Wolbachia* in *A. vulgare*, are not clear. Thierry Rigaud (1997) suggests that distributions and prevalence levels may be a consequence of internal and external factors specific to each system and questions whether a general explanation exists. This is certainly an area that needs further work, with particular attention being paid to the presence of resistance genes, to interference effects from other feminizing factors, and to correlations between prevalences and environmental factors such as temperature, photoperiod, and, in the case of marine species, salinity.

Female-biased sex ratios and associated traits, such as the occurrence of intersexual individuals, have been reported among other types of crustaceans. In some, the sexually abnormal traits appear to be controlled by nuclear genes (e.g., see Battaglia 1963; Sassaman and Weeks 1993) or environmentally induced (Adams et al. 1987). However, cytoplasmic inheritance has been implicated in the decopod crab, *Inachus dorsettensis*, the copepod, *Tigriopus japonicus*, and the crab, *Leptomithrax longipes*, although, in each of these cases, the presence of a cytoplasmic sex-determining microorganism has not yet been demonstrated and these cases await further work (Rigaud 1997).

All known cases of feminization by endosymbionts involve crustaceans, except two. The Asian corn borer moth, *Ostrinia furnacalis*, and its close relative, the Adzuki bean borer, *Ostrinia scapulalis*, are currently the subject of some exciting research by a group led by Daisuke Kageyama. In *O. furnacalis*, Kageyama and his colleagues at the University of Tokyo first reported all-female strains in 1998. Again, examination of egg-hatch rates and mortality rates in all-female families appeared to rule out differential male mortality during the immature stages. Antibiotic treatment showed the female biases to be tetracycline-sensitive, suggesting that a bacterium was involved (Kageyama et al. 1998). Subsequently, a *Wolbachia* that is taxonomically similar to that causing feminization in *A. vulgare*, was found in feminized strains of the Asian corn borer and the Adzuki bean borer, and is a candidate for being the causal agent (Kageyama et al. 2000). Intriguingly, not all the all-female strains tested positive for *Wolbachia* and Kageyama and his colleagues believe that factors other than *Wolbachia* are involved in the production of all-female broods in these species of moth. It would be interesting to discover whether any maternally inherited nuclear genes, possibly originating as a translocation of part of the *Wolbachia* genome, caused feminization in these species. If so, this would parallel the suggested evolution of a nuclear feminizing factor, the *f* factor, in *A. vulgare* (see p. 126).

The parallel with the feminizer of *A. vulgare* extends further. As in the woodlouse infected with *Wolbachia*, antibiotic treatment of feminizer lines of the Asian corn borer led to a strong male bias in progeny of cured females. These moths are ancestrally female heterogametic, with the result that feminized ZZ males, when cured of the bacterium and mated with normal (ZZ) males, can only produce male offspring. It is perhaps no coincidence that in the cases of *Wolbachia*-induced feminization, the host is ancestrally female heterogametic and I expect further instances of *Wolbachia*-induced feminization to be discovered in the Lepidoptera and other insect taxa with heterogametic females in the near future. Laurence Hurst (1993) lists a variety of other arthropod species that may harbor cytoplasmic sex factors. These include several coccids, an aphid (Hemiptera), and one butterfly. This list would certainly be a useful starting point for future investigations. Outside the Arthropoda, there is also some evidence that female tissue may be under cytoplasmic control in the hermaphroditic nematode *Caenorhabditis elegans* (Sulston and Hodgkin 1988).

Sex Determination Systems of Hosts Susceptible to Feminization

That microbe-induced feminization is apparently so much commoner in crustaceans than in any other taxonomic group leads to the following question: Is sex determination more easily compromised in the crustaceans than in other groups and if so, why? Thierry Rigaud (1997) contrasts the known cases of feminization with those of another manipulative effect of inherited symbionts on their hosts, CI. Noting that a variety of microorganisms can cause feminization, while just a single group of organisms, the *Wolbachia*, are known to cause CI, despite the greater taxonomic diversity of the hosts that suffer CI, he concludes that the fundamental basis of feminization must largely be a property of the host that is feminized rather than the feminizer. This interpretation is perhaps weakened both by the subsequent discovery of *Wolbachia*-induced feminization in the Asian corn borer and Adzuki bean borer moths and by a consideration of male-killing endosymbionts (see chapter 6) in which there is considerable diversity in both the symbionts causing male-killing and their hosts. In the case of male-killing, work on ladybird beetles (Coccinellidae) has shown that the success of a symbiont results from interactions involving both the ability of the symbiont to manipulate its host's reproduction and the tractability of the host to such manipulation. The latter is, in turn, a consequence of the both hosts' genetic systems and their ecological circumstances.

This is not to say that Thierry Rigaud's conclusion is fatally flawed. Sex determination systems within the Crustacea are certainly highly variable, with both male and female heterogametry occurring and the environment also having a considerable influence in many species. In *A. vulgare*, females are heterogametric, but both temperature and photoperiod can also influence sex determination. In uninfected *O. gammarellus*, males are heterogametic, while in *G. duebeni* a polygenic sex determination system exists, with the expression of the genes, and so the sex produced, also being affected by photoperiod. It thus seems likely that the factors that control sex determination in the Crustacea are highly flexible and may be prone to manipulation. Yet, while the ease with which a host's sex determination system can be manipulated is undoubtedly of importance, the abilities of the symbionts should not be ignored. The need for flexibility on the part of the symbiont is perhaps gleaned from the fact that the feminizers in some Crustacea, and in the only two known insect cases, are *Wolbachia*, which is well known for its flexibility in the ways that it can manipulate its host's reproduction. It would then perhaps be more accurate to say that the fundamental basis of feminization is dependent on both the flexibility of the symbiont's manipulation of its host and the susceptibility of the host's sex determination system to interference (Majerus 1999).

This summation is endorsed by the remarkable findings of Tetsuhiko Sasaki and his colleagues. Investigating the types of manipulation employed by *Wolbachia*, Sasaki micro-injected the feminizing *Wolbachia* of *O. scapulalis* into embryos of the Mediterranean flour moth, *Ephestia kuehniella*. This moth is known to harbor CI *Wolbachia*, but can be freed from infection by tetracycline treatment. When *O. scapulalis Wolbachia* were successfully injected into embryos of CI-cured *E. kuehniella* lines, expectation was that either the recipient strains would show feminization, indicating that the *Wolbachia* was in control of the phenotype it produced, or that CI would result, leading to the deduction that the phenotype expressed by *Wolbachia* was a function of its host. The resulting moths did produced female-biased sex ratios. However, breeding experiments showed that the female bias was due to neither some function of CI, nor feminization, but to male-killing (Sasaki *et al.* 2000). Here, then, the same *Wolbachia* strain has different phenotypic effects on different species of hosts from the same order of insects.

How Are Genetic Male Crustacea Feminized?

In *A. vulgare*, whether an embryo develops into a male or a female appears to depend on the activity of just one single gene. This gene blocks the expression of one or more other genes that cause development of a gland called the androgenic gland. In essence, this is a male-determining gland, for its main function is to produce male sex hormones. If its formation is prevented, a female adult results. Because the difference between male and female development is so simple in this species, there exists a fairly simple pathway of feminization that could be employed by an inherited symbiont: all it need do is inhibit androgenic gland development. The exact molecular mechanism employed by *Wolbachia* to suppress androgenic gland formation is not yet clear, but the bacterium appears to mimic the action of what might be called "female genes," carried on the W chromosome and so absent from males, that prevent androgenic gland development in normal (WZ) females. It is too early to say whether the "female genes" in normal woodlice and the feminizing genes in *Wolbachia* work in the same way at a molecular level, or simply appear mimetic in producing the same final result. It is perfectly possible that the two sets of genes may cause disruption at different stages in the development of the androgenic gland development.

The simplicity and malleability of sex determination mechanics in crustaceans can be deduced from a consideration of the types of sex chromosomes found in isopods. Sex determination here may depend on XX/XY, XX/XO, or ZZ/ZW systems, with examples of both XX/XY and ZZ/ZW systems being found in species of the same genus (Table 5.1).

If, then, the crustacean sex determination system is the reason for the ease with which it may be corrupted, with the result that the Crustacea are a "hot spot" for feminizing microorganisms, we should ask why other groups are not as susceptible. The obvious answer is that the sex determination systems of most other organisms are more complex and so not as easily corrupted. In *Drosophila*, for example, despite enormous amounts of research, involving the rearing of billions upon billions of flies in laboratories around the world, no

TABLE 5.1
Examples of various types of chromosomal sex determination found in isopods. (Adapted from Juchault and Rigaud 1995.)

	Sex Chromosomes of Each Sex (Males Given First)		
Suborder	*XO/XX*	*XY/XX*	*ZZ/WZ*
Asellota		*Asellus aquaticus*	*Jaera marina*
Flabellifera	*Tecticeps japonicus*		*Dynamene bidentata*
Valvifera			*Idothea balthica*
Oniscidea		*Porcellio dilatatus dilatatus* *Armadillium nasatum* *Helleris brevicornis*	*Porcellio dilatatus petiti, P. rathkei, P. laevis* *Armadillium vulgare* *Eluma purpurascens, Oniscus asellus*

simple genetic change has been discovered that will transform an XY male into a viable and reproductively functional female. Investigations into the mechanism of sex determination in *Drosophila* indicate that several factors have to operate in concert to fully feminize a chromosomal male (p. 56). The task of mimicking the multipart sex-determining mechanism in insects would obviously be harder for potential feminizers than it is in Crustacea. The rarity of endosymbiotic feminizers in insects (only two confirmed cases) may simply be a function of this difficulty.

The Dual Feminizing Action of Wolbachia

Having worked on one evolutionary system (the blackening of certain moths in industrial regions) for 40 years and another for over 20 years (all things sexual in certain ladybirds), I have come to dual conclusions. First, when one has a nice, clean answer to a specific question, the best thing to do is to stop work on the system as quickly as possible, for if one delves any further, the clean answer will usually end up getting very muddied. Second, if one looks at a particular organism closely enough, examining it from every angle and investigating as many aspects of its biology and ecology as possible, the intimate understanding of the organism gained will throw up a range of interesting problems that one would never otherwise have identified. The researchers will be rewarded because they will be armed with the knowledge needed to address these problems. To my mind, it is to the great credit of the French groups who have conducted so much of the work on feminizing *Wolbachia* of woodlice, that they did not abandon the system once they had identified the general mechanism of feminization in this system. The reward for their tenacity was the discovery that these *Wolbachia* have a second feminizing mechanism, independent of that affecting androgenic gland development.

The second mechanism of feminization in woodlice was discovered when *Wolbachia* from infected lines were injected into adult males with fully differentiated and functional androgenic glands. The result was the gradual feminization of these males, with an intersex phenotype being produced, suggesting that feminization did not simply involve the prevention of the formation of the androgenic gland (Juchault and Legrand 1985). Amazingly, further tests showed that in these individuals the androgenic gland remained active and continued to produce androgenic hormone. This hormone, when injected into immature (i.e., non–sexually differentiated) individuals, caused them to develop as males, showing that the hormone remained active. Yet the hormone was ineffective in its *Wolbachia*-bearing producer, suggesting that the *Wolbachia* in some way interfere with the hormone's receptors in target tissues.

Sex Determination Warfare

Why *Wolbachia* have two independent mechanisms that affect the sex determination of their hosts is not known, but it is a question that certainly warrants attention because it almost certainly results from a host/symbiont arms race. As discussed in chapter 4, due to differences in patterns of inheritance, biparentally inherited nuclear genes are in conflict with uniparentally inherited cytoplasmic genes. Theoretically, this conflict should lead to selection favoring the production of more male than female progeny by uninfected females, who increase their long-term fitness (thought of as the number of grandchildren they produce) simply by producing the rarer sex (Fisher 1930). Alternatively, selection may favor nuclear genes that attack the cytoplasmic feminizer, thereby reducing its feminizing effect either directly or indirectly. Direct action may involve reducing the *Wolbachia*'s replication rate, thereby reducing the proportion of eggs to which it is vertically transmitted, or killing it. Indirect action may involve suppressing the feminizing effect of the *Wolbachia*, or development of novel means of male development in the host, thereby bypassing *Wolbachia's* mechanism of androgenic gland inhibition.

Some experimental evidence supports both direct and indirect responses. Breeding experiments on woodlice, in which those females that produced most males were selected as parents for each subsequent generation, resulted in an increase in the proportion of males produced. Genetic analysis showed that this response to selection was dependent on autosomal genes that increased the proportion of males in progeny (Rigaud and Juchault 1992).

Before considering these suppressor genes in detail, the full picture of feminization in *A. vulgare*, as far as it is currently known, must be explained. In fact, there are not one, but two distinct feminizing factors known in *A. vulgare*. These are the *Wolbachia* discussed above (following the French researches, notated as *F*) and a DNA sequence, which I shall call a gene, probably derived from *F*, that has been unstably incorporated into the nuclear genome of *A. vulgare*. This nuclear feminizer is named *f*. Both factors feminize ZZ males and both are maternally inherited (although paternal inheritance of *f* has been observed on rare occasions). Despite their similarity, there are some differences between the two feminizing factors. For example, some ZZ + F females produce a high proportion of intersexual progeny. These intersexes are of three types: (a) female intersexes (iF) (Figure 5.2), which are reproductively capable females with some male external characteristics; (b) male intersexes (iM) (Figure 5.3), which are sterile but have male external features; and (c) functional males with female genital openings (Mfg) (Figure 5.4 is a normal male, shown for comparison). Both iF and iM harbor *Wolbachia* and begin to develop as males, but the male development then stops so that further development is as females. Mfg individuals begin to develop as females but then male development takes over. The *f* factor does not lead to the production of intersexes.

A third factor that affects sex determination in *A. vulgare*, is a masculinizing gene (*M*). This is a dominant autosomal gene. In females with the ancestral ZW sex chromosomes, *M* simply changes females into males (some Mfg intersexes are sometimes produced), overriding the female-determining factors on the W chromosome. The effect of *M* in the presence of a feminizing factor depends on which factor it is confronting (Box 5.1). In *f*-carrying strains, *M* restores males,

Figure 5.2 Micrographs of the undersides of female intersex *Armadillium vulgare* (cf. Figure 5.4). Note the very reduced copulatory brushes and the lack of copulatory brushes on the legs. (Courtesy of Robert Comte and Gilbert Martin, UMR-CNRS 6556, Génétique et Biologie des Populations de Crustacés, Poitiers, France.)

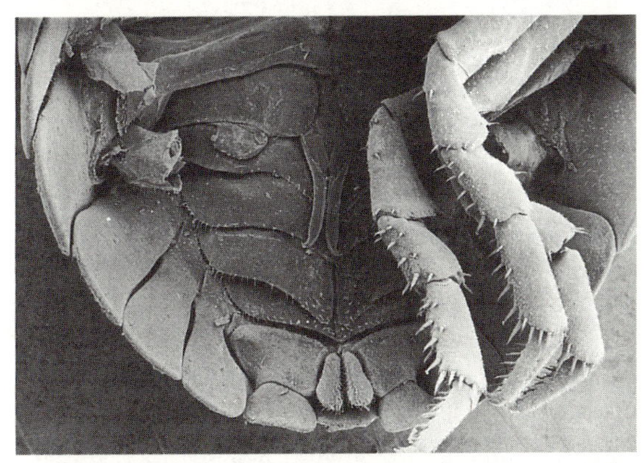

Figure 5.3 Micrographs of the undersides of male intersex *Armadillium vulgare* (cf. Figure 5.4). Note the reduced male appendages and the lack of copulatory brushes on the legs. (Courtesy of Robert Comte and Gilbert Martin, UMR-CNRS 6556, Génétique et Biologie des Populations de Crustacés, Poitiers, France.)

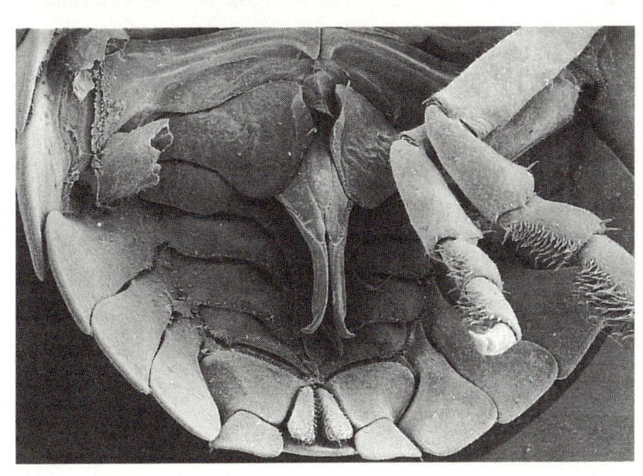

Figure 5.4 Micrographs of the undersides of normal male *Armadillium vulgare*. Note the large paired male appendages and the copulatory brushes on the legs. (Courtesy of Robert Comte and Gilbert Martin, UMR-CNRS 6556, Génétique et Biologie des Populations de Crustacés, Poitiers, France.

Box 5.1 The effects of the masculinizing factor *M* on the various feminizing factors in *Armadillium vulgare*. The feminizing *Wolbachia* is denoted as B, the feminizing factor by *f*, and the masculinizing dominant allele as *M* with the nonmasculinizing recessive allele as *m*. *M* is inherited in a Mendelian fashion. The feminizing factors are here assumed to be transmitted to all progeny. (Derived from Rigaud 1997.)

Cross	Progeny
1. Male (ZZ *Mm*) × Female WZ *mm*	→ 1/4 Male (ZZ *mm*);
	1/4 Male (ZZ *Mm*);
	1/4 Male (WZ *Mm*);
	1/4 Female (WZ *mm*)
2. Male (ZZ *Mm*) × Female ZZ *mm* + *f*	→ 1/2 Male (ZZ *Mm* + *f*);
	1/2 Female (ZZ *mm* + *f*)
3. Male (ZZ *Mm*) × Female WZ *mm* + B	→ 1/2 Female (ZZ *mm* + B);
	1/2 Female or intersex (ZZ *Mm* + B)

M restores masculinity in the presence of *f*, but not in the presence of B.

nullifying the effect of the feminizer. Although *M* blocks the phenotypic expression of *f*, it does not interfere with its vertical transmission, and the proportion of *f* carrying progeny is roughly the same, irrespective of whether *M* is present or not.

By contrast, when faced with *F*, *M* is unable to disrupt the feminizing action of the symbiont and the number of reproductive males produced does not increase. The few males and Mfg individuals produced in *F* strains that were induced by crossing experiments to also carry *M* were the product of inefficient transmission of the *Wolbachia* rather than the effect of *M*. However, it would not be true to say that *M* has no effect on the phenotypes of *F* carriers, for it does seem to increase the proportion of intersexes (both iF and iM). The interactions between *M* and the two feminizers mean that a range of different genotypes and phenotypes with respect to sex occur in *A. vulgare*.

Detailed biochemical and cytological analysis showed how *M* interacts with *F* and allows speculation on the reason that *F* has two modes of feminization. Although, the *M* gene is not able to override the feminizing *Wolbachia*, it does counteract the primary feminizing mechanism of the bacterium, the inhibition of androgenic gland development, as manifest in initial sexual development producing male characteristics. This initial masculinization is stopped at an early stage by the secondary feminizing effect of the *Wolbachia*, and further development is of female traits as a result of inactivation of androgenic hormone receptors. It thus seems likely that the secondary feminizing effect of *Wolbachia* has evolved as a result of an arms race, being a response to the evolution of the *M* gene (or other resistance genes), which in turn evolved as a response to the primary feminizing effect of the *Wolbachia* and/or factor *f* (Rigaud 1997). The two feminizing effects of *Wolbachia* in *A. vulgare*, and the evolution of *M*, provide an elegant set of genetic products of an evolutionary arms race.

Conflict between Feminizers: No Honor among Thieves

The question remains; what happens when *F* and *f* occur in the same population? Theoretically, as models of Bull (1983) and Taylor (1990) have shown, one would expect that the *Wolbachia* with a transmission efficiency of around 80–90% would outcompete the *f* factor, which has a lower vertical transmission efficiency of around 60–65%. However, studies of a population of *A. vulgare* at Niort in France, over a 24-year period, have shown that this expectation is precisely the reverse of what has been observed. In the Niort population, *f* has increased in frequency at the expense of *F*. The reason appears to depend, at least in part, on the fact that individuals carrying *f*, but being male because they also harbor *M*, suppressing the feminizing action of *f*, can transmit the *f* factor to progeny. Males can never transmit *F* to progeny. Particularly in populations with a considerable female bias, the transmission of *f* through males will give it a great advantage over *F*, despite its lower transmission through females. It is interesting to note that the frequency of *M* increased substantially in the Niort population over the study period. Furthermore, *M* does not occur in wild populations unless feminizers are present. This is to be expected because in normal populations with a 1:1 sex ratio, a masculinizing gene would be selected against due to the negative frequency-dependent selection inherent in Fisher's sex ratio theory.

A model of the possible series of events in a population of woodlice newly invaded by a feminizing *Wolbachia* has been suggested by Juchault et al. (1993). Following invasion, *F* will begin to spread. Should a part of the *F* genome which included the primary feminizing genes (but not the secondary genes, p. 126), be susceptible to transfer into the nuclear genome of the host, for example, if it were a jumping gene, this would generate *f*. The evolution of resistance genes (including *M*) in the host against the feminizing effects of *F* would put pressure on *F* to respond, with the result that the secondary feminizing action of the *Wolbachia* develops. The pressure of resistance, and the fact that *M*-carrying males can now transmit *f* but not *F*, causes the replacement of the *Wolbachia* by the feminizing sequence, now incorporated into the host's genome, that it has given rise to.

This specific plan, based on what is known about sex determination in *A. vulgare*, may be slightly extended to give a model of the evolution of sex chromosomes. In populations in which males are ZZ and females are ZZ+*f*, all the chromosomes, with the exception of the feminizer sequence, are pairs. If the

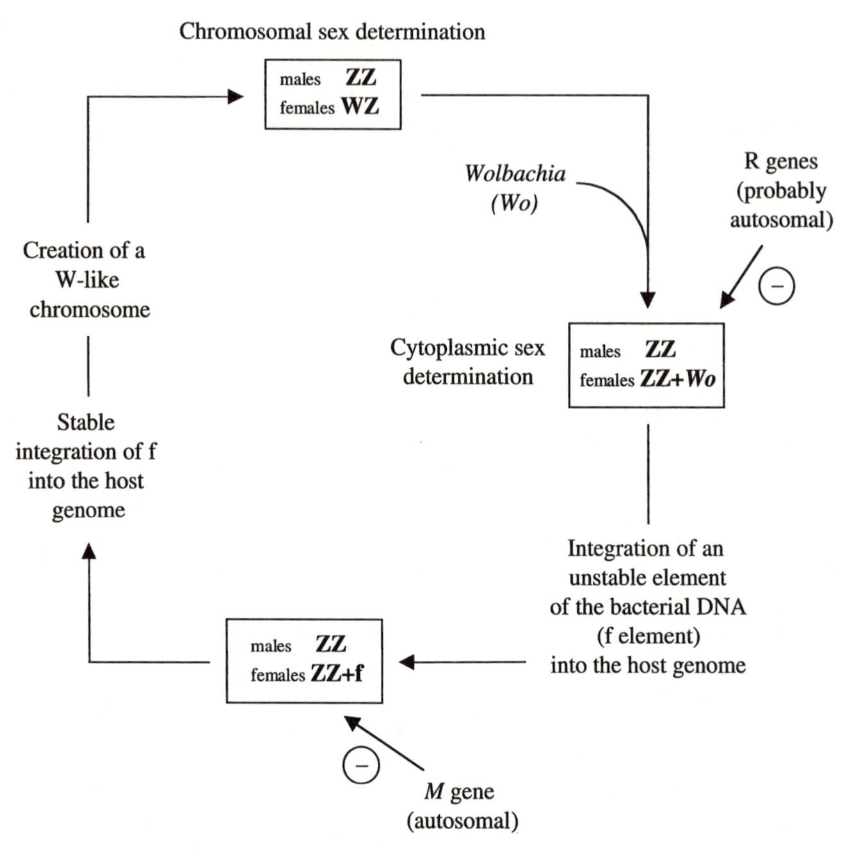

Figure 5.5 A simple model of the evolution of sex chromosomes inferred from studies of *Armadillium vulgare*. (Adapted from Rigaud 1997.)

feminizer sequence inserts into, or fuses onto the end of, any existing chromosome, then this chromosome and its homologue will effectively have become a new pair of sex chromosomes. All that is needed in the *A. vulgare* case is to stabilize the integration of *f* into the nuclear genome. Once this has been achieved, the new chromosome produced will effectively be a W-like sex chromosome containing the genetic material that determines femaleness. This simple model of sex chromosome evolution is shown schematically in Figure 5.5.

To me, the competition between *F* and *f*, and the outcome in favor of the latter, is fascinating for two reasons. First, it provides another natural example of suicide. In effect, when *F* transfers its feminizing ability to the nuclear genome of its host, it is, in the long term, committing suicide. Second, this case again shows that the gene (or DNA sequence) is an important unit of selection. Jumping genes, otherwise known as transposable elements, are genetic law breakers. These elements have the ability to move around the genome. If the *f* factor is a transposable element, as has been suggested, its interaction with *M* and its success over *F* illustrates how this sequence, rather than the feminizing phenotype is selectively favored.

Feminizing microorganisms manipulate the sex of their hosts, directly favoring females, the sex through which they can be inherited. Their action does not cause the death of their carriers. The ultraselfish symbionts of many other invertebrates are less benign. These microbes assassinate male hosts. This cutthroat strategy is called male-killing and is the subject of the next chapter.

CHAPTER 6

Male-Killers

> A species . . . has a great will to survive. . . . So babies were born: the girl babies lived, the boy babies died.
>
> —John Wyndham, *Consider Her Ways*

SUMMARY

Due to their maternal inheritance, parasitic intracellular microorganisms have no interests in being in males, since they cannot be vertically transmitted by male hosts. However, three types of advantage may be gained as a result of killing male hosts. First, the male-killer may be transmitted horizontally from the corpse of dead males. Second, female siblings of dead males that carry the male-killer due to shared inheritance will be less likely to inbreed. Third, resources that would have been consumed by males may be released for consumption by female siblings. These different strategies lead to differences in the timing of male host pathogenicity. The aim of this chapter is to introduce the protists and the diverse array of bacteria that have adopted a male-killing strategy.

The evolutionary rationale underlying male-killing and the behavioral and ecological features of host species that may make them liable to male-killer invasion are explained with particular reference to mosquitoes, ladybird beetles, and nymphalid butterflies.

TWO TYPES OF MALE-KILLER

The successful invasion of an organism by a feminizing microorganism appears to depend, at least partly, on the malleability of the host's sex determination system. Conversely, the second mechanism by which inherited microorganisms distort host sex ratios in favor of females, the killing of males, appears to depend more on the ecology of hosts.

Male-killers are known in a wide range of insect species, being reported among beetles, butterflies, moths, true bugs, flies, and parasitic wasps. Cases of male-killing in other groups of invertebrates certainly await discovery.

The basic rationale of male-killers is the same as that of other microbes that are inherited in the cytoplasm of cells. If transmission to the next generation is exclusively through the cytoplasm of eggs, bacteria in a female gamete, which after fertilization becomes destined to develop into a male, are stuck. They cannot be passed to the next generation, so they would appear to be up an evolutionary creek.

How may bacteria escape this "dead-end" predicament? Two of the options described in chapter 5 (p. 116) concern us here. The first is to replicate as fast as possible to produce large numbers of copies, killing male hosts in the process, and then to migrate out of the male host's decaying corpse. Here we are suggesting a degree of horizontal transmission from males to females. The second option is to kill male hosts and in so doing preferentially favor female hosts, which carry clonally identical copies of the male-killer. This, then, involves kin selection: bacteria in males sacrificing themselves for the benefit of their relatives in female hosts. Both these strategies have been realized.

The Timing of Male Death

Laurence Hurst, in a thorough analysis of male-killers, recognized that the two different strategies—one involving horizontal transmission of the sex ratio distorter, the other conferring advantages on infected females by the death of males—should cause the death of males to occur at different times (Hurst 1991). To increase the chances of successful horizontal transmission, male-killing should occur late in host development, thus allowing the microbe to multiply within its host so that many individuals would be released after host death. Conversely, if the death of male hosts leads to the death of the male-killers within those host individuals, but favors infected females, death early in host development will be favored if the benefit to females involves infected females gaining resources of one sort or another that would otherwise have been used by males. We therefore have two categories of male-killers—late male-killers and early male-killers (Table 6.1)—and the two can be discussed separately.

TABLE 6.1
Types of male-killing caused by cytoplasmic endosymbionts. (Proposed by L. D. Hurst 1991.)

Type	*Hosts*	*Causative Agents*	*Rationale*
Late male-killing	Mosquitoes	Microsporidia	Males killed in final larval instar promoting horizontal transmission from host corpse.
Early male-killing	Various insects	Various bacteria	Males killed in embryogenesis to promote resource reallocation or reduce inbreeding. Horizontal transmission rare.

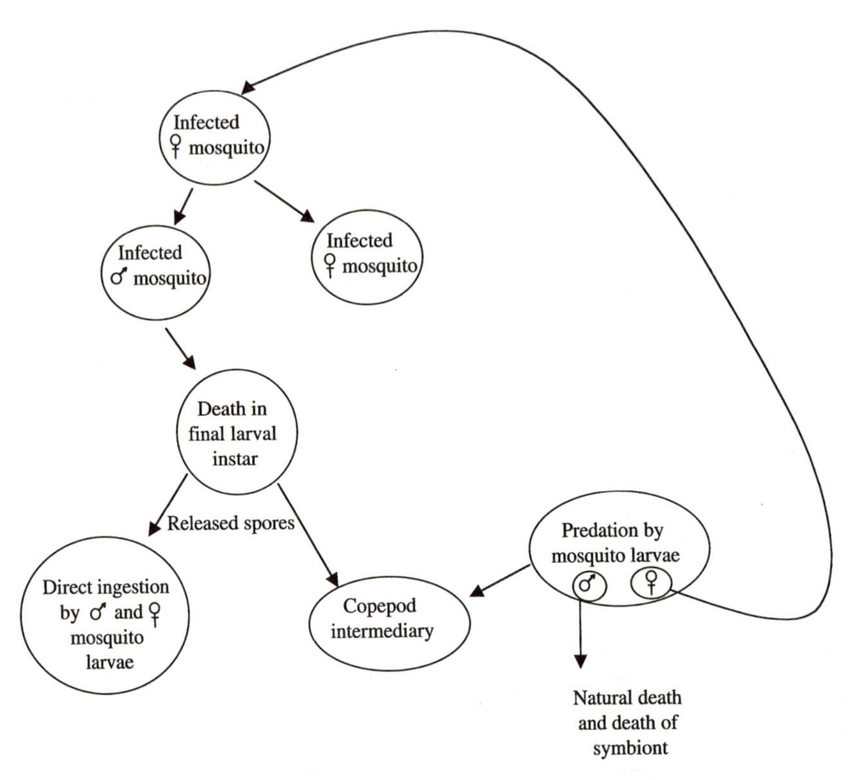

Figure 6.1 Modes of transmission of late male-killing Microsporidia in mosquitoes.

Late Male-Killers

Confirmed cases of late male-killing are currently known only from one group of insects, the mosquitoes, and involve a single group of microorganisms, the Microsporidia, some of which we have already encountered feminizing amphipods (p. 120). Over two dozen species of mosquito are known to be infected with male-killing Microsporidia. The Microsporidia have two modes of transmission (Figure 6.1). Those in female mosquitoes can be vertically transmitted to offspring. Those in males, for which vertical transmission is not an option, can be horizontally transmitted to other hosts either via larval cannibalism or when spores are released into the water following the death of their host, which typically occurs when the larva is in its final instar. Released spores may be taken in by other mosquito larvae by direct ingestion or within a copepod intermediary, which may become infected by the spores and subsequently be preyed on by the mosquito larvae (Bechnel and Sweeney 1990). Either male or female larvae can be infected and neither shows ill effects following novel infection, at least within their own lives. Indeed, the Microsporidia appear to have no effect at all on newly infected males. Such males complete their life cycle normally and, since the symbionts cannot be vertically transmitted from adult males, they die when their host dies. Newly infected female hosts also complete their life cycle normally, but these pass the microsporidian into their eggs. The males in this following generation may be killed, while females continue to transmit the symbionts vertically.

This is the simplest pattern for a late male-killing microsporidian. However, in nature, the situation is often more complex and infection by microsporidians does not always follow this path. Some microsporidians cause the death of both host sexes, while others kill neither.

These cases beg several questions. How do the microsporidians kill their hosts? How do some selectively kill male rather than female hosts? Why do host deaths typically occur precisely in the fourth larval instar? What is the reason for the variation in the strategies employed by different microsporidians, some being pathogenic to both sexes, some just to males, and some to neither sex? Not all of these questions have clear answers. In particular, the methods by which microsporidians determine the sex of their host and by which they cause their host's death await discovery.

Hurst (1991) argued that the timing of host death is crucial if the death of the host is coupled with horizontal transmission. By causing death in the fourth larval instar, the microsporidian is maximizing its transmission rate. He suggests, justifiably, that horizontal transmission needs to occur in water. Further, he argues that the pupal case, the last aquatic stage of the life cycle, might act as a barrier to symbiont release. In consequence, he states that causing host death in the final larval instar, but not before, maximizes the number of microsporidian spores released into the water to be taken up by copepods that would then vector the microsporidian if preyed on by other mosquito larvae.

Horizontal versus Vertical Transmission

The range of strategies employed by different microsporidians may arise out of differences in the probability of vertical and horizontal transmission, in particular ecological situations (Hurst 1991). In bodies of water in which copepods are common, horizontal transmission is likely to be efficient, in which case lethality of both host sexes might be expected, particularly if vertical transmission efficiency through eggs is poor, as it often is. Conversely, if copepods are scarce, making horizontal transmission a hazardous affair, selection is likely to favor vertical transmission through eggs from female hosts and horizontal transmission out of the corpses of killed male hosts. Why some microsporidians are benign to both sexes remains unclear. Three possibilities exist. First, such a strategy might be beneficial if vertical transmission via eggs in females were coupled with some degree of sexual transmission from males during copulation. Second, if copepods or other vectors were scarce or absent, making horizontal transmission via an intermediary highly improbable, the best strategy for a microsporidian in a male host might be to leave it alive so it had a chance of succumbing to cannibalism and being horizontally transmitted that way. Third, males might carry a gene that suppressed the pathological effects of the microsporidian. Clearly, in this economically important group of flies, further experimental work is needed to examine these possibilities and investigate both the mode of host-killing and how some Microsporidia are able to kill selectively with respect to the sex of their hosts.

Recent work by Sayaka Morimoto has revealed another probable case of late male-killing, in the Oriental tea tortrix moth, *Homona magnanima* (Morimoto et al. 2001). In SR families, both male and female larvae hatch, but males show slower larval development than females, and then develop a blotchy appearance before dying in their second or third instars. Although *Wolbachia* has been found in this moth, *Wolbachia* presence is not correlated with the SR trait. Rather, the causative agent is thought to be a virus. Homogenate of moths from SR lineages were passed through a series of micro-filters before being injected into uninfected individuals. The micro-filters were of a size that should have allowed passage of viruses, but excluded larger microbes, such as bacteria or protists. If a virus is confirmed as the causative agent of male-killing in this moth, it will broaden the range of symbionts that produce this phenomenon. Furthermore, as many pathogenic viruses of Lepidoptera are known to horizontally transfer out of the decaying corpses of their victims, it seems likely that, here too, late-male-killing is associated with high levels of horizontal transmission.

Early Male-Killing

Fitness Compensation

If a male-killing symbiont has little or no chance of escaping from its doomed host by transmitting horizontally, some other advantage to that symbiont must result from the death of males if the symbiont is to persist and even spread. This must happen to compensate for the fitness loss due to the death of the sons of infected females; otherwise, the male-killer will fail to spread and will eventually be lost. Two mechanisms of fitness compensation have been suggested. The first results from the adverse effects of reproducing with very close relatives, such as full siblings (p. 6). In the context of male-killing, daughters of females infected with a male-killer that has high vertical transmission efficiency cannot mate with full siblings, as all their brothers will be dead. Thus, an advantage to male-killing may result from the avoidance of inbreeding and consequent fitness losses through inbreeding depression. Although this advantage is generally listed under early male-killing, in fact, unless there is also competition for limited resources, it matters little when males are killed, as long as it is before males reach reproductive maturity.

This is not the case for the second mechanism of fitness compensation. The death of males may be advantageous to females if resources that would have been used by the males become available to infected females. This mechanism is called *resource reallocation*. The resources involved in the known cases of early male-killing are most commonly, but not exclusively, nutrients. As nutrients used by infected males before they die will be wasted and not be available to females, the earlier the males die, the greater the advantage gained by the females. Here, then, there will be selection for male-killing symbionts to kill as early as possible.

The rationales for these two potential advantages to early male-killing are the same as those for reduction in inbreeding and reallocation of resources from pollen to seeds in mitochondrially induced CMS in hermaphroditic plants (p. 93).

In the context of advantages to male-killing as a result of resource reallocation, it is imperative that the resources made available to females from the death of males become preferentially available to females carrying the same male-killer, if the male-killer is to spread. From the symbiont's point of view, killing males is only a viable strategy if a kin-selective benefit results (see Hamilton 1964). Death of males will benefit infected females when male and female progeny from one mother occur close together. Because some groups of organisms give birth to many progeny in a small area, such as those insects that lay eggs in tightly packed batches, while others spread their offspring over much larger areas, the distribution of male-killers is unlikely to be random in the invertebrates. Groups of species with particular behavioral or ecological similarities will be prone to early male-killer infection. Other groups lacking such traits should be more or less immune to these male-killers, simply because there will be insufficient fitness compensation from male-killing to allow for the invasion or evolution of this strategy.

Taxa Affected by Early Male-Killers

Early male-killing microbes have been recorded in a taxonomically diverse array of hosts in five orders of insects and in two species of mite, with a variety of different bacterial taxa having been reported as agents (Table 6.2). Within these groups, as expected, a nonrandom pattern of male-killer distribution has been observed (Majerus and Hurst 1996). Hot spots for male-killing are known in the milkweed bugs (Hemiptera), nymphalid butterflies, particularly of the genus *Acraea*, and the ladybird beetles (Coccinellidae). As a group, the ladybird beetles have been most extensively studied and have been proposed as a

TABLE 6.2
The diversity of early male-killers in different orders of insects.

Order	*Male-Killer*
Hemiptera	Unknown bacteria
Lepidoptera	*Wolbachia*, Group VI *Spiroplasma*
Coleoptera	*Wolbachia, Rickettsia*, γ-proteobacteria, Flavobacteria, Group VI *Spiroplasma*
Hymenoptera	γ-proteobacteria
Diptera	*Wolbachia*, Group II *Spiroplasma*

model system for the study of many aspects of early male-killing (Majerus and Hurst 1997). Nearly half the confirmed cases of early male-killers occur in ladybird beetles, and it is within this group that the evolutionary reasons for the spread of male-killing are most clearly understood. It is thus perhaps worth detailing the ladybird male-killing story first to use as a model against which other instances of male-killing can be compared.

Early Male-Killing in Ladybird Beetles

The First Ladybird Male-Killer

The Russian geneticist, J. J. Lusis, first reported that strongly female-biased sex ratios were produced by some female two-spot ladybird beetles, *Adalia bipunctata*, in 1947. The achievements of this Russian, who worked in Russia throughout the Stalinist purges and the era of Trofim Lysenko's influence, when Mendelian genetics was not considered politically attractive, should not be underrated. The great Russian/American evolutionary geneticist Theodius Dobzhansky first brought the subject of ladybirds and the problems that their color pattern variation posed, to Lusis's attention, before Dobzhansky emigrated to America. Lusis took up Dobzhansky's challenge and continued to work on the two-spot ladybird for over 40 years, many times having to resort to conducting his genetic experiments in the secrecy of the attic of his house (I. A. Zakharov, pers. comm.).

Lusis (1947) noted that the tendency of some females to produce mainly daughters was maternally inherited and that the proportion of females of this type varied geographically. In 1991, with my student Greg Hurst, I began to investigate all female broods of *A. bipunctata* in England. We rapidly established female-biased (SR) lines and normal (N) lines from a Surrey sample. Observations showed that, in the SR lines, the proportion of eggs that hatched was generally about half that in the N lines. This suggested that the primary sex ratio of the SR lines was normal, but that the secondary sex ratio was distorted because male embryos died before hatching from their eggs (Plate 2a). Genetic analysis of crosses showed that the SR trait was inherited only from mothers, and not from the males occasionally produced by SR females.

If maternally inherited, the agent causing the male-killing seemed likely to be a mitochondrial gene, a bacterium, a protist, or a virus. Testing the easiest of these first, antibiotic treatment (tetracycline) administered in golden syrup (e.g., Plate 2b), effected a cure. Eggs produced three or four days after treatment began showed elevated hatch rates and a higher proportion of males produced, eventually rising to about 50%.

Once the causative agent was known, light and electron microscopy of hemocytes revealed a bacterium in the cytoplasm of SR females, but not in N females or in males, with differences in the distribution and densities of bacteria being seen in different host tissues (Figure 6.2, Plates 3a and 3b). The bacte-

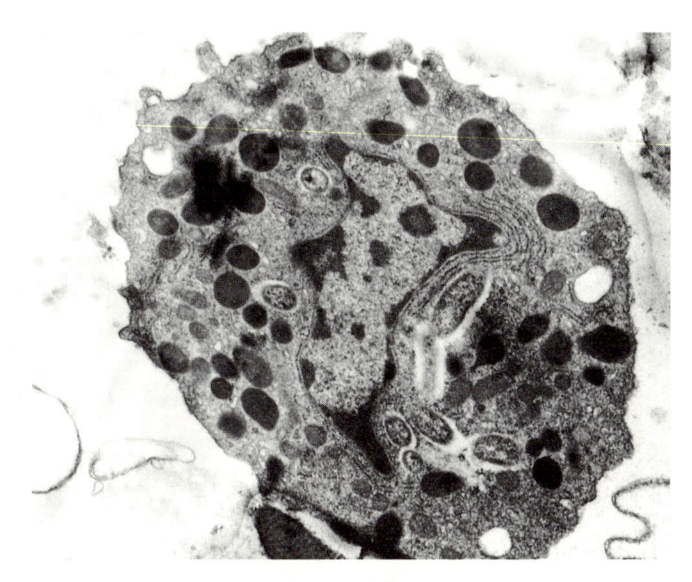

Figure 6.2 A transmission electron micrograph of a hemocyte from a female *Adalia bipunctata* infected with a male-killing *Rickettsia*. The bacteria are circular or oval, speckled gray, and surrounded by a vacuole. (Courtesy of Brij Gupta and Greg Hurst.)

rium was then identified as an α-proteobacterium of the genus *Rickettsia*, with closest homology to *Rickettsia typhi* and *Rickettsia prowaskii*. *Rickettsia typhi* is the organism that causes cattle scrub fever, while *R. prowaskii* is a pathogen of humans, more commonly known as murine typhus (Werren et al. 1994).

The Diversity of Male-Killers in Ladybird Beetles

Collaboration with Professor Ilia Zakharov from the Vavilov Institute of General Genetics in Moscow allowed us access to samples from across Russia, including some of the locations where Lusis obtained his original "strains without males." The first samples supplied by Zakharov, from Moscow and St. Petersburg, revealed a number of SR females. However, the characteristics of infection in these Russian beetles were subtly different from those of the British SR lines. In particular, the vertical transmission efficiency of the trait in Russian lines seemed to be higher than in British lines, manifest in a lower number of males produced in SR families. Molecular identification of the symbiont in the Russian lines showed that a different bacterium, related to group VI *Spiroplasmas*, was present. The story does not end there, however; ultimately, two more male-killing bacteria have been identified from *A. bipunctata*. The most recent two are different strains of *Wolbachia* (Hurst et al. 1999). Their discovery was of considerable interest, for these were the first discovered cases of *Wolbachia* causing male-killing in their host.

TABLE 6.3
Egg-hatch rates and progenic sex ratios of six lines showing the male-killing trait, from a sample of 36 *Adalia bipunctata* females collected in Moscow. Other females produced normal sex ratios. The symbionts identified from these lines are given. (From Majerus et al. 2000.)

Line	*Egg Hatch Rate*	*Number of Progeny*	*Sex Ratio (Proportion Male)*	*Symbiont*
Mos 3	0.478	34	0.147	*Rickettsia*
Mos 6	0.422	44	0.114	*Wolbachia* "Z"
Mos 9	0.391	33	0	*Spiroplasma*
Mos 18	0.444	43	0	*Wolbachia* "Y"
Mos 33	0.511	35	0	*Spiroplasma*
Mos 35	0.388	18	0	*Spiroplasma*

One piece of analysis concerning this plethora of bacterial male-killers in *A. bipunctata* is of particular note. A small sample supplied by Zakharov, who collected them from the side of a building in Moscow as they were moving to their overwintering sites, was used to produce 36 matrilines. Of these, six proved to be SR, with all four of the male-killers identified from the two-spot being present (Table 6.3) (Majerus et al. 2000). For reasons that I will return to later (p. 183), the finding of four male-killers coexisting in a single population of their hosts was extraordinary because the observation of such coexistence is in direct conflict with theory.

Following the discovery of male-killing bacteria in *A. bipunctata*, samples of other ladybirds were bred to see whether they also produced biased sex ratios. Female-biased sex ratios had already been reported in some other species of ladybird and some of these were investigated first. The shocking-pink, long-legged ladybird, *Coleomegilla maculata*, from America, was found to carry a different type of male-killing bacterium, a Flavobacterium, somewhat similar to the mutualistic gut bacteria of cockroaches and some termites. An Asian ladybird, *Harmonia axyridis*, was found to harbor a *Spiroplasma* similar to that found in *A. bipunctata*. Another Asian species, *Cheilomenes* (=*Menochilus*) *sexmaculatus*, was found to bear yet another type of male-killer, a γ-proteobacterium that is somewhat similar to the causative agent of the plague, *Yersinia pestis* (Majerus 2001). Tests on ladybirds in which female-biased sex ratios had not previously been reported brought forth four other cases. Adonis' ladybird, *Adonia variegata*, from southern Europe, and *Propylea japonica*, from Japan, were both found to harbor Flavobacteria similar to those in *C. maculatus*. Ten-spot ladybirds, *Adalia decempunctata*, from Germany, and *Coccinula sinensis*, from Japan, were found to carry *Rickettsias*. In addition, there are records of female-biased sex ratios in three other species of coccinellid that have not yet been subjected to detailed analysis (see Table 6.4).

TABLE 6.4
Details of known cases of male-killing in ladybird beetles.
(Adapted and amended from Majerus 1999.)

Host Species	*Countries*	*Agent*	*Prevalence*	*Evidence**
Adalia bipunctata	England, Scotland, Holland, Denmark, Luxembourg, Belgium, France, Germany, Poland, Russia, Kyrgyzia	*Rickettsia*	0.01–0.11	f-bsr; lhr; mi; as; hs; m; sDNA; PCR
Adalia bipunctata	Germany, Russia, Sweden	*Spiroplasma* (Group VI)	0.03–0.13	f-bsr; lhr; mi; as; hs; m; sDNA; PCR
Adalia bipunctata	Russia, Sweden	Two strains of *Wolbachia*	each < 0.1	f-bsr; lhr; mi; as; hs; m; sDNA; PCR
Adalia 10-punctata	Germany, England	*Rickettsia*		f-bsr; lhr; mi; as; hs; m; sDNA; PCR
Harmonia axyridis	Japan, Russia, South Korea	*Spiroplasma* (Group VI)	0.02–0.86	f-bsr; lhr; mi; as; hs; m; sDNA; PCR
Coleomegilla maculata	U.S.	Flavobacterium	0.23	f-bsr; lhr; mi; as; sDNA; PCR
Adonia variegata	Turkey, England	Flavobacterium	0.13	f-bsr; lhr; mi; as; sDNA; PCR
Hippodamia 5-signata	U.S.	Unknown		f-bsr; lhr; mi
Cheilomenes 6-maculatus	Japan	γ-proteobacterium	0.13	f-bsr; lhr; mi; as; PCR
Coccinula sinensis	Japan	Flavobacterium	0.23	f-bsr; lhr; mi; as; PCR
Propylea japonica	Japan	*Rickettsia*	0.07–0.26	f-bsr; lhr; mi; as; PCR
Coccinella 7-punctata	England	Unknown	< 0.01	f-bsr; lhr
Calvia 14-guttata	England, Canada	Unknown	< 0.05	f-bsr; lhr, mi

*Evidence given as: f-bsr = female-biased sex ratio; lhr = low egg hatch rate in infected lines; mi = maternal inheritence; as = antibiotic sensitive; hs = heat sensitive; m = microscopy; sDNA = DNA sequencing; PCR = detection of symbiont using symbiont-specific PCR reaction.

The Principal Advantage to Male-Killing in Ladybirds: Resource Reallocation

Although male-killers have not been found in all species analyzed, the ladybirds do seem to be very prone to invasion by bacterial male-killers. So what features of ladybirds make them so prone to attack? To answer this question, we have to return to the rationale behind why male-killers kill males, committing suicide in the process.

As mentioned earlier, a male-killer in a male host is at an evolutionary dead end if it cannot be transmitted horizontally from the corpse of its host, since it is not vertically transmitted in sperm. Experiments feeding infected eggs and

Figure 6.3 A tightly clustered ladybird egg clutch.

larvae to uninfected larvae and adults have failed to produce novel infections, so it appears that horizontal transmission via ingestion of infected tissue is, at best, a very rare event. However, although a male-killer in a male host appears to be at a dead end by killing its male host, its life need not be worthless if it increases the survival of other hosts that harbor the same male-killer. And in certain types of ladybird, the ladybirds that feed on aphids, this is exactly what happens.

Three aspects of the biology of aphidophagous ladybirds have to be understood to comprehend why these are so prone to male-killer infection. First, aphidophagous ladybirds lay eggs in tight clutches (Figure 6.3). Second, aphids are prone to rapid population increases and crashes so that they are a highly ephemeral prey for ladybirds. Third, ladybirds are highly cannibalistic. These facts are not independent (see Majerus and Majerus 1997 for review).

The cannibalistic tendencies of ladybirds may be a consequence of the ephemerality of their prey. If aphid populations suddenly crash because of the activities of their many parasitoids and predators, including ladybirds, or because of a deterioration in their own food supply, ladybird adults and larvae will be left starving. Cannibalism is then likely to evolve because those that develop this habit will be more likely to survive than those that do not develop cannibalism. Both adult and larval ladybirds are cannibalistic (e.g., Plate 2c and 2d). First instar larvae, for instance, habitually eat any unhatched eggs in their clutch as soon as their mouth parts have hardened sufficiently to pierce the shell of another egg (Figure 6.4). They will eat eggs that fail to hatch because the eggs were not fertilized, those that fail to hatch because the embryo has died for some undetermined reason, and those that while developing have not yet hatched. This last category is crucial because it imposes a strong selective pres-

Figure 6.4 Neonate ladybird larvae begin to consume unhatched eggs in their clutch as soon as their mandibles have hardened sufficiently.

sure for larvae to develop very rapidly and get out of the egg as quickly as possible: slow developers will simply become a meal for their more rapidly developing siblings. The result is that neonate ladybird larvae are very small when they hatch, often much smaller than the prey that they are going to have to try to catch and subdue (Figure 6.5). The small size of ladybird larvae when they disperse from their egg clutch means that they are usually too poorly resourced to survive long unless they find a meal. Experiments have shown that normal *A. bipunctata* larvae must find food within about 24 hours of dispersing from their eggs or they starve to death. Not only must they be able to find food, but they must also be able to catch it and hold on to it. Many fail on one or another of these counts, so that the mortality rate of neonate larvae is often very high (Banks 1955, 1956). Indeed, Wratten (1976), working in southern England, found that up to 50% of neonate *A. bipunctata* larvae died without catching their first aphid meal.

When Suicide Is Painless

The reason why these ladybirds are so prone to male-killers now comes sharply into focus. Consider two clutches of eggs, one from an N female and one from an SR female. Allow both to have normal fertility (>90%), to have a 1:1 primary sex ratio, and to comprise 20 eggs. In the normal clutch, almost all are destined to hatch, and probably 18 or 19 will, with just 1 or 2 slow developers or infertile eggs falling foul of their siblings (e.g., Plate 2e). These will then disperse and will have to find food within about a day and the prey will have to be small enough for them to manage. In the SR clutch, half the eggs fail to hatch because they are destined to be male but carry a male-killer. The 9 or 10 larvae that do hatch are all females. Each of these, on average, has the contents

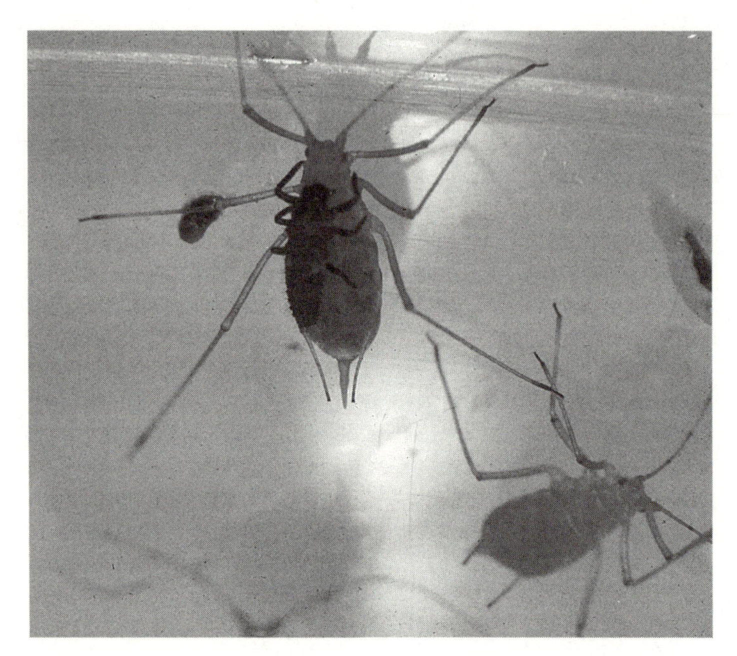

Figure 6.5 Neonate ladybird larvae are frequently smaller than the aphids that they prey on. This larva has successfully climbed onto an aphid and is feeding with its mandibles imbedded.

of 1 unhatched dead male egg to consume (Plate 2f). This consumption cannot be called cannibalism in a scientific sense, because the definition of cannibalism is the killing and eating of one's own species, so we will call this *sibling egg consumption*. The soma of the dead male eggs is highly nutritious for these small female larvae. Indeed, it contained when it was laid, exactly the nutrients that the female larva has used to develop to hatching point. If all these resources are available to the newly hatched female larvae, these will gain a considerable survival advantage over female larvae from a normal clutch, most of whom do not get this large extra meal. By their sacrificial suicide, the bacteria in male eggs thereby increase the fitness of clonally identical copies of themselves in female siblings of their hosts.

Male-Killing and Kin Selection

Here, then, we have an exquisite and extreme case of kin selection in which one bacterium lays down its life for its closest kin. Professor John Maynard Smith (1964) first coined the term *kin selection* following analysis by Hamilton (1964) into the importance of genetic relatedness for the evolution of altruistic behaviors. Hamilton's analysis, which built on an idea suggested by Fisher (1930) with respect to the evolution of aposematic coloration and calculations of levels of genetic relatedness by Haldane (1955), led to Hamilton's rule. This rule states that an altruistic gene will spread if the cost it imposes on its carrier (C) is less than the benefit (B) that accrues to the recipient of the altruistic behavior that it gives rise to multiplied by the genetic relatedness between the altruist and the beneficiary (r). This is frequently written in the form that if $rB-C > 0$, an altruism gene will evolve. Here, then, selection is not maximizing the fitness of a particular individual in terms of the number of its direct descendants. Rather, it is maximizing the fitness of all descendants, whether direct or otherwise, weighted by the likelihood of these descendants sharing the same genes as a result of their relatedness. Professor J.B.S. Haldane once wrote that he would jump into the rapids and give up his life if this would save two of his brothers from drowning (because they would share approximately half of his genes), or four half-brothers (each with a quarter of his genes), or eight cousins (carrying 12.5% of his genes).

Kin selection lies behind much unusual and apparently altruistic behavior in the animal kingdom. Young eastern ground hornbills help their parents in rearing subsequent broods, rather than attempting to breed themselves. Related male lions will cooperate to take and retain control of a pride, even if only one to them mates with the lionesses in the pride. Workers of many social insect species selflessly forgo reproduction in favor of other tasks that benefit the colony. In all these instances, the individuals are still behaving in such a way as to maximize the passage of identical copies of the genes they carry into the next generation. In the case of the male-killing bacteria, the conditions are more in favor of the evolution of kin selection than in the above cases, for here individuals of the same clone are, for all intents and purposes, genetically identical, so the relatedness (r) is 1.

Of course, if the male embryo begins to develop, it will begin to use up these resources. The longer the embryo survives before being killed by the male-killer, the more the resources that will be used up to no good effect. So here we have a reason why male-killing bacteria in coccinellids appear to kill males very early indeed: the earlier they can identify and kill male hosts, the greater the level of nutrients in the soma of their hosts' eggs that will be available to their clonal relatives' hosts.

The advantage that accrues to female larvae of male-killed clutches through the consumption of unhatched eggs has been assessed in a number of ways and in a number of species. Evidence suggests that in *A. bipunctata*, a neonate larva that eats one male-killed egg, in the absence of other food or water, survives, on average, half as long again before starving to death, compared to a larva denied such an extra meal. Similar figures have been found for other species. In the small ladybird, *Coccinula sinensis*, which lays unusually large eggs for its size, the longevity advantage is considerably greater, with the average survival of a larva from a male-killed clutch being nearly double that of a larva from a normal clutch (Figure 6.6). Further, *A. bipunctata* larvae from male-killed clutches were found to be larger when they dispersed from their natal clutch, were able to subdue a greater size range of aphids, and traveled farther in search of aphids following dispersal, before dying from starvation in the ab-

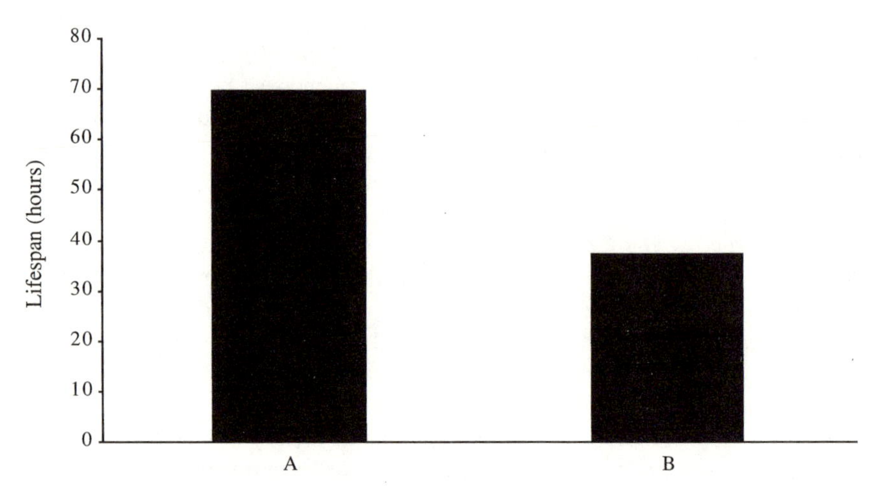

Figure 6.6 Survival of neonate larvae of *Coccinula sinensis*, deprived of water or aphid prey. A) Average survival of larvae allowed to consume a single male-killed egg. B) Average survival of larvae denied an egg meal.

sence of such prey. The greater resources consumed by SR larvae before they dispersed from their egg clutches also lead to more rapid first instar larval development and higher likelihood of survival to first ecdysis, particularly at low aphid densities (see Majerus 1994 for review). Similar advantages resulting from resource reallocation via sibling egg consumption have recently been shown in four other coccinellids bearing male-killers, *H. axyridis*, *P. japonica*, *C. sexmaculatus*, and *C. sinensis* (Majerus 2001). The advantage to male-killing through resource reallocation due to sibling egg consumption thus seems to apply generally among aphid-eating ladybirds.

Avoidance of Cannibalism

Interestingly, there is a second advantage to male-killing that is dependent on the cannibalistic behavior of neonate ladybird beetle larvae. The synchronicity of hatching of the larvae from a clutch of eggs is not exact. The result is that a small proportion of developing embryos that would hatch, given time, are attacked and eaten by their siblings simply because they are slow developers. Again, we can imagine two clutches of eggs, one from an SR female and one from an N female, and again both with 20 eggs. Let us now allow that 1 male and 1 female embryo in each clutch are slow developers, so that they run the risk of being cannibalized by their early-hatching siblings. In the N clutch, this is a very strong likelihood because, assuming full fertilization, there will be 18 early-hatching larvae and only 2 as-yet-unhatched eggs available. In the SR clutch, the situation is different for two reasons. First, there will be only 9 early-hatching larvae (the 9 female embryos that develop quickly), so there will be fewer potential cannibalizers. Second, there will be 11 unhatched eggs to attack, the slow-developing female embryo and 10 dead male eggs. On many occasions this may give the slow-developing female additional time to get out of her egg because, just through random choice, the already hatched larvae may attack other unhatched eggs before they get to her. This is even more likely to be the case if the early-hatching larvae preferentially attack the most resource-rich eggs available, since these will be the male eggs that have been killed early: in the slow-developing female eggs, the embryo will have used up much of the resource.

Evidence that this reduction in cannibalism does indeed lead to an increase in the number of female larvae that hatch in male-killed compared to normal clutches has been obtained for three species of coccinellid, *A. bipunctata*, *H. axyridis*, and *P. japonica*. By removing larvae from egg clutches as they hatched using a single-bristle brush, larvae were prevented from interfering with any of the other eggs. This treatment led to a significant increase in the egg-hatch rate of uninfected clutches, but not of infected clutches. The deduction is that a significantly greater proportion of females die as a result of sibling cannibalism in uninfected clutches than in infected clutches.

A third possible advantage to male-killing dependent on the cannibalistic behavior of ladybird larvae has yet to be verified by empirical evidence. Not only do larvae consume any unhatched eggs in their clutch before dispersal, they will also eat conspecific larvae once they have dispersed, particularly if other manageable prey are scarce. In interactions between two larvae, assuming that neither is restricted because it is shedding its skin or pupating, the larger larva usually wins and eats the smaller one (Majerus 1994). Since larvae from SR clutches are on average larger than N larvae when they disperse, the SR larvae are likely to gain another cannibalistic advantage when clutches of eggs are laid close together and at the same time by SR and N mothers.

The cannibalistic behavior of ladybirds, coupled with their habit of laying their eggs in batches, underlies their susceptibility to male-killing bacteria. In this case, the advantage from male-killing results primarily from the redistribution of resources from the killed males to their sisters, which have a high probability of carrying the same male-killer lineage of bacteria. The other potential advantages of male-killing—providing conditions for horizontal transmission or the avoidance of inbreeding—seem to have little importance here.

The Population Dynamics of Male-Killing in Ladybird Beetles

The high potential levels of fitness compensation that have been demonstrated in aphid-feeding ladybirds should allow male-killers to invade and spread. The question that then needs to be asked is this: will the male-killer spread to such an extent that the host population is driven extinct as a result of lack of males? From observation, the answer appears to be *no*, for the recorded prevalences of male-killers in ladybirds are not inordinately high, usually ranging from 1% to about 50% of females. Only in samples of *H. axyridis* have prevalences of specific male-killer strains higher than 50% been recorded (p. 169).

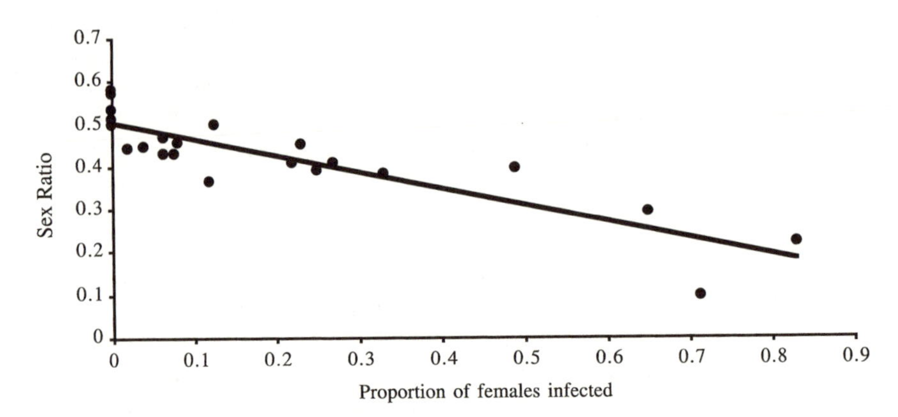

Figure 6.7 The correlation between the population sex ratio and male-killer prevalence in different populations of *Harmonia axyridis*. (Courtesy of Tamsin Majerus.)

The prevalence of a male-killer should influence the population sex ratio of its host: the higher the prevalence, the greater the female bias that results. This does appear to be the case. In *H. axyridis*, assessments of the proportion of *Spiroplasma*-infected females in samples collected at a variety of locations in Asia have shown a close correlation to the proportion of females in the populations from which the samples were drawn (Majerus et al. 1998; Majerus 2001) (Figure 6.7). While this correlation is encouraging, fitness compensation and male-killer prevalence are not the only factors that need to be considered in determining the behavior of a male-killer in a wild host population.

To understand the dynamics of the spread of a male-killer, two other factors have to be considered: the vertical transmission of the trait and any cost that the bacterium imposes on female hosts.

Vertical Transmission Efficiency

The vertical transmission efficiencies of inherited microorganisms are very variable. As we have already seen, vertical transmission of CI *Wolbachia* is frequently near perfect (>99%) in laboratory cultures, but somewhat lower in the field (95–97%). Feminizing *Wolbachia* in *Armadillium vulgare* are transmitted to about 90% of host offspring. From laboratory studies, it appears that the levels of vertical transmission of male-killing bacteria of ladybirds are similarly variable. This is seen in the differences in vertical transmission efficiencies of the four male-killers of *A. bipunctata* from Moscow (Table 6.5). Highest vertical transmission has been recorded for the male-killing *Spiroplasma* of *H. axyridis* (from Sapporo) and *A. bipunctata* (from Moscow) (Majerus et al. 1999, 2000), at in excess of 99%. However, transmission efficiencies of other male-killers are lower, typically in the range 80–95%. This means that in each generation, infected females give rise to some males.

The production of some male progeny by infected females may be an impor-

TABLE 6.5
Vertical transmission efficiencies from six matrilines bearing male-killing bacteria. The matrilines lines are derived from the families shown in Table 6.3. (From Majerus et al. 2000.)

Line	*Number of Families*	*Total Progeny*	*Sex Ratio (Proportion Males)*	*Mean Vertical Transmission Efficiency ± s.e.*
Mos 3 (*Rickettsia*)	5	123	0.220	0.719 ± 0.0166
Mos 6 (*Wolbachia* Z)	6	222	0.126	0.856 ± 0.0017
Mos 9 (*Spiroplasma*)	6	130	0.015	0.984 ± 0.0006
Mos 18 (*Wolbachia* Y)	7	168	0.012	0.988 ± 0.0004
Mos 33 (*Spiroplasma*)	4	170	0.006	0.994 ± 0.0001
Mos 35 (*Spiroplasma*)	5	199	0	1

tant feature in the maintenance of the host population, because males from infected females will gain the same resource advantage as their infected sisters. Thus, these males will have, on average, a greater chance of survival than males from uninfected mothers. As a result of this fitness difference between males from infected and uninfected mothers, the system as a whole presents an interesting and complex picture. Selection acting on the male-killer will select for efficient male-killing (and low costs on female hosts). However, selection acting on male hosts will act to reduce the vertical transmission efficiency of the male-killer.

Work on the ten-spot ladybird, *Adalia decempunctata*, and *H. axyridis*, has indicated that male-killing efficiency is dependent on the density of bacteria in developing male embryos. In the ten-spot ladybird, females subjected to high temperature, which causes a reduction in bacterial density, produced a higher proportion of males, some of which were subsequently shown to be infected by the bacterium but at a very low density. In *H. axyridis*, very young infected females sometimes initially produce a normal or only marginally female-biased sex ratio. The clutches giving rise to these progeny have a high hatch rate. The low egg-hatch rates and all-female progeny characteristic of male-killing only arise after three or four egg clutches have been laid. Interestingly, in ladybirds showing this progressive sex ratio trait, the timing of male death also appears to be related to bacterial density. In the first clutches to show low hatch rates, the proportion of those eggs that do not hatch that are gray, that is, show signs of embryonic development, is very high, but this proportion declines in later clutches in favor of yellow eggs (Table 6.6, Plates 3e–h). These findings sug-

TABLE 6.6
Temporal changes in egg-hatch rates and sex of progeny from a single *Spiroplasma*-infected female of *Harmonia axyridis* mated repeatedly to a single male. Yellow eggs show no signs of development. Gray eggs show signs of embryonic development, but do not hatch. Initially the hatch rate is high, and some males are produced. The proportion of eggs that hatch gradually declines and after week 7 drops below 30%, suggesting that some female eggs are perishing. (Data given in week blocks, starting one week after eclosion.)

Week	*Proportion of Eggs Yellow*	*Proportion of Eggs Gray*	*Proportion of Eggs That Hatched*	*Number of Male Progeny*	*Number of Female Progeny*
1	0.03	0.08	0.88	11	23
2	0.11	0.23	0.66	5	22
3	0.17	0.38	0.46	1	29
4	0.38	0.19	0.43	0	25
5	0.43	0.08	0.49	0	19
6	0.47	0.11	0.41	1	28
7	0.42	0.18	0.40	0	21
8	0.46	0.25	0.28	0	16
9	0.51	0.41	0.08	0	11
10	0.47	0.49	0.03	0	7

gest that a reduction in the vertical transmission efficiency of the bacterium will lead to an increase in surviving males in infected clutches. One only has to imagine that the interaction between host and symbiont involves genes of the host that influence the proportion of an infected female's progeny that receive the bacterium to see that those genes that reduce vertical transmission will be selected for in male progeny that are both free of the bacterium and gain from the resource reallocation resulting from the death of their infected brothers. In addition, such progeny do not suffer other costs of bearing the bacterium (see p. 152). The interrelationships between bacterial density, vertical transmission, and host genes that influence vertical transmission are complicated, and empirical evidence from ladybirds is urgently needed to understand these more fully.

Despite this, there is no doubt that the vertical transmission efficiencies of the various male-killers in coccinellids do differ. These differences, in themselves, raise some interesting questions. For example, are the different vertical transmission efficiencies seen in different systems a result of characteristics of the male-killer, or its host, or the interaction between the two? As yet, the jury is out. However, some observations do shed a little light. First, the male-killers with the highest vertical transmission efficiencies recorded from ladybirds are both *Spiroplasmas*, which might suggest that efficient transmission is a feature of this bacterium. Second, the differences seen in the vertical transmission efficiencies of the four male-killers discovered from Muscovite *A. bipunctata* and particularly the differences between the two strains of *Wolbachia* would suggest that the host alone is not critical in determining male-killer transmission efficiency. Possibly the replication characteristics of the different types of bacteria that have evolved male-killing play a crucial role in determining their vertical transmission efficiency, but this may be mediated through selection acting on their hosts (particularly male hosts) and selection acting at the population level (p. 187).

One final point is worth making. Virtually all information on the vertical transmission efficiencies of male-killing bacteria has come from laboratory investigations. Given that temperature, antibiotics, host dormancy, and host density can all affect bacterial transmission (Hoffmann and Turelli 1997), field vertical transmission efficiencies may be considerably different from those measured in captive ladybird beetles. Again, this matter warrants urgent attention.

Horizontal Transmission

Late male-killing is associated with the horizontal transmission of the symbiont from the corpses of dead males. Here horizontal transmission is a crucial aspect of the generation upon generation dynamics of the male-killer. This is not the case for early male-killers. Various attempts have been made to transmit male-killers of coccinellids horizontally within or between species by natural means. These have included feeding infected eggs to uninfected larvae or adults, or using parasitoids or parasites as vectors. As yet, all such attempts have failed. It thus seems likely that if horizontal transmission does occur, its frequency is so low that it will have a negligible impact on the prevalence of infection. However, very rare instances of horizontal transmission may be important in the context of novel invasion of a population or a species.

A number of very similar male-killers are found in different species of ladybirds. *Adalia bipunctata* and *Adalia decempunctata* harbor almost identical strains of *Rickettsia*. The DNA sequence of the 16S rDNA gene of the *Spiroplasma* that infects *H. axyridis*, is virtually the same as that infecting the two-spot ladybird. In both these instances, the relatedness of the male-killers appears closer than that of the hosts, suggesting that horizontal transmission between species has occurred. In the latter case, artificial horizontal transmission of the *H. axyridis* male-killer into *A. bipunctata* has been shown to produce a male-killing phenotype. If horizontal transmission between species has occurred, we may ask how such transfer may have operated. Several possibilities exist. Apart from the possibilities of transfer via larval consumption of eggs, or vectoring of male-killers via parasitoids or parasites, interspecific hybrid matings also occur occasionally (Plates 3c and 3d). Although these matings are not successful, it is possible that male-killing bacteria could be transmitted during copulation. Only one transmission event needs to be successful to found a novel infection. Therefore, while horizontal transmission of early male-killers is probably of little importance over ecological time, it may be crucial over evolutionary time.

The Cost of Male-Killing in Ladybird Beetles

The likelihood that a male-killer will invade, spread, and be maintained in a particular host depends not only on the vertical transmission of the trait, but also on the fitness of females infected with the symbiont compared to uninfected females. The latter will, in turn, be the product of, on the one hand, the advantage accruing to infected females from resource reallocation or reductions in inbreeding and, on the other hand, any cost associated with carrying the trait. Here we are considering costs in addition to that incurred through the loss of male progeny.

In five species of ladybird with three different male-killers (*A. bipunctata* and *P. japonica–Rickettsia*; *H. axyridis–Spiroplasma*; *A. variegata* and *C. sinensis*–Flavobacterium), costs in terms of decreased rates of egg laying, higher levels of infertility, and shorter adult life span have been reported (e.g., Matsuka et al. 1975; Hurst et al. 1994). Similar negative effects have been reported in other insects that harbor male-killers (Table 6.7).

In addition, some lines of *Spiroplasma*-infected *H. axyridis* have been observed to give rise to hatch rates considerably lower than 50%. One explanation of this observation is that some female embryos are also killed by the *Spiroplasma* (Table 6.6). Analysis of the fluctuations in the proportion of yellow and gray unhatched eggs produced by SR females through their lives suggests that bacterial density affects pathogenicity. In eggs produced by very young females, bacterial density is low enough that a few males survive, and those that are killed die late in embryo development. As bacterial density increases, the egg-hatch rate pattern typical of male-killing is seen, with half hatch rates, and most unhatched eggs being yellow. Eventually, as the female ages, bacterial density passes a second threshold that is sufficient to cause the death of female embryos late in their development within eggs. This is manifest in a decline in the egg-hatch rates to well below 50%, with the proportion of gray eggs again increasing. Formal demonstration of whether the late-killed embryos in clutches from old mothers are females awaits the development of a reliable test of embryo sex for this species.

As yet, the painstaking experiments to assess the costs or benefits to female ladybirds of carrying male-killing bacteria have only been undertaken in five species and all these experiments have been conducted in the laboratory. Given that it is not in a male-killer's interests to impose a cost on hosts that are able to transmit them (i.e., female hosts), selection might be expected to promote mutualism between male-killing bacteria and female hosts. To this end it is to be hoped that cost/benefit analysis of harboring male-killers is conducted on other ladybird/male-killer systems. In the five studied cases, costs to female hosts, rather than benefits, have been found. It may be that here, although it would be in the selective interests of the bacterium to become harmless or even beneficial to female hosts, metabolic limitations and the need to proliferate to ensure transmission limit its evolutionary potential in this direction.

TABLE 6.7
Species in which negative effects of male-killing have been observed.

Adult Female Longevity	*Fecundity*	*Fertility*	*Female Embryo Survival*
Adalia bipunctata	*Adalia bipunctata*	*Drosophila willistoni*	*Drosophila willistoni*
Harmonia axyridis	*Harmonia axyridis*		*Harmonia axyridis*
Propylea japonica	*Cheilomenes sexmaculatus*		
	Coccinula sinensis		
	Propylea japonica		
	Spodoptera littoralis		
	Epiphyas postvittana		

Male-Killing in Ladybirds: Testing the Predictions

The advantages to male-killing in ladybird beetles seem clear, with specific ecological traits promoting male-killing behavior by inherited endosymbionts. The advantages of sibling egg consumption and avoidance of cannibalism allow predictions to be made about which species of coccinellid should be prone to male-killer invasion and which should not be. Species that do not lay their eggs in clutches, do not indulge in sibling egg consumption/cannibalism, and are not limited by ephemeral food (i.e., do not have aphids as their primary prey), are less likely to have male-killers than those with these features. In a survey to test these predictions, 30 coccinellid species were assayed for the presence of male-killers. Of these, 20 were deemed to have ecologies and behaviors that would promote male-killing (e.g., Plates 4a–c). The other 10 each lacked one or more of the factors necessary for male-killer spread (e.g., Plates 4d–f). Table 6.8 shows that male-killers were found in about half of those species predicted to be liable to male-killer invasion. However, in none of the species lacking one or more of these traits has a male-killer been found.

MALE-KILLERS AND RESOURCE REALLOCATION IN OTHER INSECTS

The evolutionary rationale that leads to male-killers being so widespread among the aphid-eating ladybirds is now well understood, at least qualitatively. Certainly there are still many questions that need to be addressed before we can say we understand why, for example, male-killers vary in their vertical transmission efficiency or why there is variation in the prevalence of different male-killers in host populations of the same species. However, these are quantitative questions and they are unlikely to undermine resource reallocation as the principal advantage to male-killing in the ladybirds.

In most other species in which early male-killers have been recorded, things are far less clear, although circumstantial evidence based on a consideration of

TABLE 6.8
Male-killing in ladybird beetles: testing the predictions of criteria proposed to favor male-killing. Samples of 30 species of ladybird beetle with different foods or oviposition characteristics were assayed for male-killer presence using phenotypic indicators (egg-hatch rate and progenic sex ratios). A minimum of 30 matrilines were assayed for each species. Male- killing was only detected in species that fulfilled all criteria outlined in the text (p. 153).

Male-Killer Detected In:	*Male-Killer Not Detected In:*
10 species with all criteria stipulated to make coccinellids liable to male-killer invasion	10 species with all criteria stipulated to make coccinellids liable to male-killer invasion
	10 species lacking one or more of the criteria stipulated to make coccinellids liable to male-killer invasion. Comprising: 1. Two mycophagous species (not food limited, high neonate larval survival) 2. Two vegetarian species (not food limited, high neonate larval survival) 3. Four species that lay eggs singly or in groups of two or three–three of these are coccidophagous; one feeds on adelgids 4. Two species that eat aphids as a secondary prey to coccids and are probably not heavily food limited

the known ecologies of the host insects is suggestive of a resource advantage to male-killing in a few cases. In the milkweed bug, *Oncopeltus fasciatus*, additional resources obtained through the consumption of male-killed eggs may provide neonate larvae with extra time to find suitable seed pods of their food-plant (Ralph 1977).

Both the jewel wasp, *Nasonia vitripennis*, and the mymarid, *Caraphractus cinctus*, a minute parasitoid wasp of water beetle eggs, suffer reduced competition between siblings within single hosts as a result of male death. This may lead to reduced mortality of remaining females and surviving females may be larger and produce more eggs (Skinner 1985; Jackson 1958).

In the beetle *Ips latidens*, females lay eggs in a tunnel system and interactions between early instar larvae are predominantly between siblings. Observations of high mortality and cannibalism in these early instars suggest that there is competition for the phloem upon which the larvae feed (Miller and Borden 1985). Male-killing will thus result in decreased competition for and lowered cannibalism of, infected females, simply as a result of a reduction in local larval density.

a) Virgin female Emperor moths, *Saturnia pavonia*, are able to attract males from up to 10km away by use of pheromones.

b) The wasp parasitoid *Dinocampus coccinellae* reproduces parthenogenetically, laying eggs in coccinellids.

c) Rare male *Dinocampus coccinellae* will court females, but females do not respond.

d) Sex in the esturine crocodile is controlled by the temperature at which eggs are incubated.

a) An egg clutch of *Adalia bipunctata* infected with a male-killing *Rickettsia*. Eggs are scored as yellow (male-killed or infertile), gray (inviable or liable to cannibalism), and clear (hatched).

b) Antibiotic treatment of male-killer–infected *Harmonia axyridis*. Tetracycline is administered in golden syrup.

c and d) Cannibalism is common among coccinellids.

c) Nonsibling egg cannibalism in *Harmonia axyridis.*

d) Cannibalism of a conspecific pupa by a larva of *Harmonia axyridis.*

e) Normal egg clutch of *Adalia bipunctata*. Note the high hatch rate and consumption of infertile and inviable eggs.

f) Egg clutch in *Spiroplasma*-infected *Harmonia axyridis*. Note the consumption of unhatched male eggs by neonate female larvae.

a and b) Differences in the distributions of male-killing *Rickettsia* in tissues of *Adalia bipunctata* revealed by DAPI staining. The bacteria show up as small bright flecks.

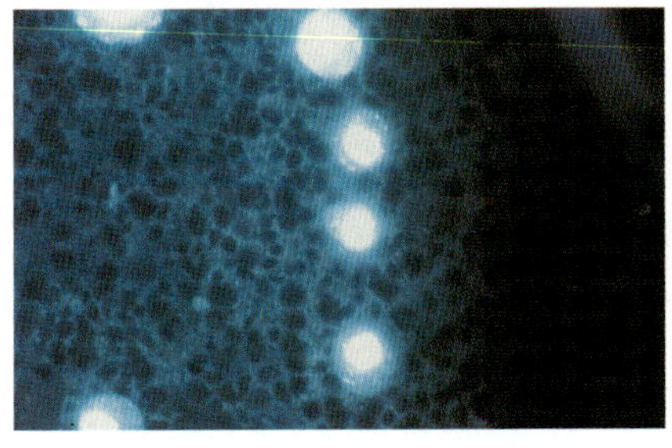

a) In hemolymph, the bacterium is restricted to the cytoplasm of hemocytes. (Courtesy of G.D.D. Hurst.)

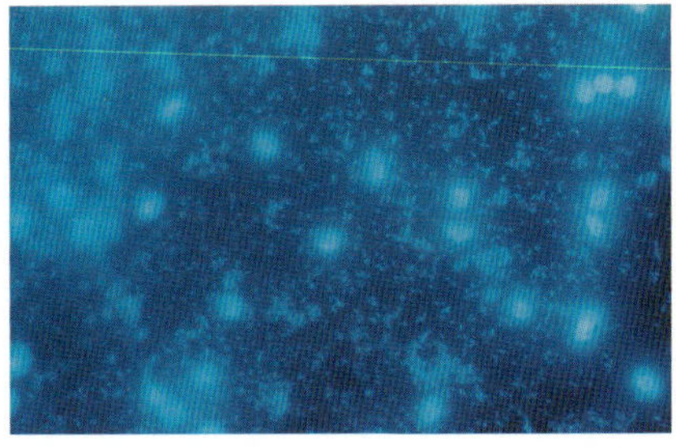

b) In immature oocytes from an ovariole, a high density of bacteria is visible. (Courtesy of J.H.G.v.d. Schulenburg.)

c and d) Interspecific hybridization in coccinellids.

c) *Adalia bipunctata* male x *Adalia decempunctata* female.

d) *Mysia oblongoguttata* male x *Coccinella magnifica* female.

e

f

g

h

e–h) Change in hatching pattern of eggs of *Harmonia axyridis* infected with a male-killing *Spiroplasma.* Clutches are shown from week 1 (e), week 5 (f), week 8 (g), and week 10 (h). Further explanation in text p. 149.

Male-killing in coccinellids. Ecological and behavioral features allow for prediction of which species of coccinellids are prone to male-killer invasion.

a) *Coleomegilla maculata.*

b) *Coccinula sinensis.*

c) *Adonia variegata.*

a–c Species bearing traits necessary for male-killer invasion, including laying eggs in clutches and feeding on aphids, and found to harbor male-killing bacteria.

d) *Chilocorus renipustulatus* (does not lay eggs in batches, feeds on coccids).

e) *Thea duovingintipunctata* (mycophagous).

f) *Henosepilachna vigintioctopunctata* (vegetarian).

d–f Species lacking trait necessary for male-killer invasion.

a) Some female *Acraea eponina*, harbor *Wolbachia* that cause a female-biased progenic sex ratio.

b) A female lekking site of *Acraea encedon* in the Botanic gardens, Entebbe, Uganda.

c) Female *Acraea encedon* flying over a lekking site on the campus of Makerere University, Kampala, Uganda. (Courtesy of F. M. Jiggins.)

d) A lekking cluster of *Acraea encedon*. (Courtesy of F. M. Jiggins.)

e) Larvae of *Acraea encedana* forage in groups until late in development.

f) Larvae of *Danaus chrysippus* are solitary through development.

PLATE 6 Male-killing in aposematic, polymorphic coccinellids

Color pattern polymorphism in aposematic coccinellids infected with male-killing bacteria.

a) *Adalia bipunctata.*

b) *Adalia decempunctata.*

c) *Harmonia axyridis.*

d) *Propylea japonica.*

e

f

g

e–g) *Cheilomenes sexmaculatus.*

Color pattern polymorphism in aposematic butterflies infected with male-killing bacteria.

a–c) Polymorphic forms of *Acraea encedon*.

a

b

c

d–f *Danaus chrysippus*.

d) Monomorphic form from Queensland, Australia.

e

f

e and f) Two of several forms occurring in East Africa. Male-killing has been detected only in populations polymorphic for color patterns. (Courtesy of F. M. Jiggins.)

a

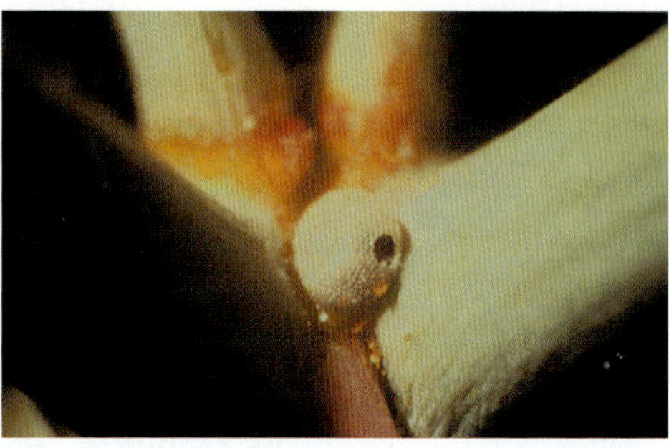

b

a and b) The habitat (a) and a host butterfly egg (b) of the egg parasitoid *Trichogramma kaykai,* which, in turn, hosts a parthenogenesis-inducing *Wolbachia*. (Courtesy of Richard Stouthamer.)

c and d) *Wolbachia* are of medical importance due to their mutualistic symbiosis with nematodes that cause disease in humans.

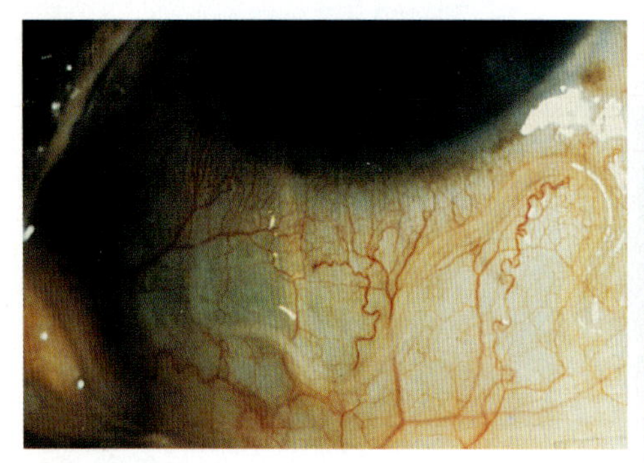

c) The infamous "eye worm," Loa loa, of West and Central Africa. The adult worm pictured is 3cm long and can traverse the eye in five minutes, causing excruciating pain. (Courtesy of Mark Taylor.)

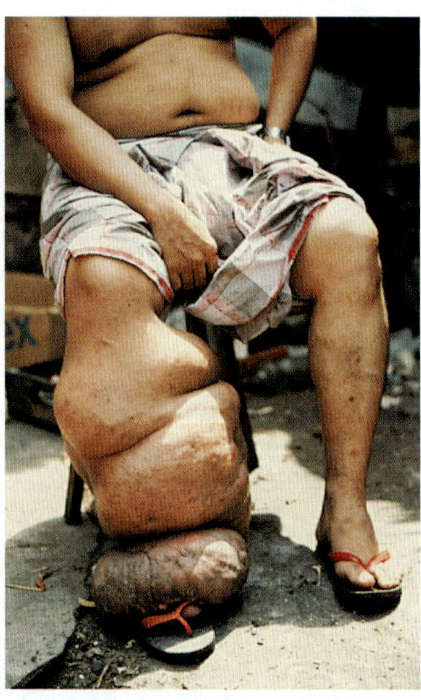

d) A Malaysian suffering from elephantiasis caused by a *Wolbachia*-harboring nematode, *Brugia malayi*. (Courtesy of Ted Bianco and Mark Taylor.)

In the tortrix moth, *Epiphyas postvittana*, competition among dispersing neonate larvae for spin-up sites, which are limited, leads to high first instar larval mortality (MacLellan 1973; Geier and Briese 1980). Again, the reduction in density resulting from male-killing may lead to increased survival of infected as opposed to uninfected females.

Empirical evidence is needed to establish increased fitness of infected females in all of these instances. In other species shown or thought to bear male-killing symbionts, the likelihood of a resource advantage accruing from male-killing has either not been ascertained or is deemed low based on ecological considerations (see Hurst and Majerus 1993 for review).

Can Inbreeding Avoidance Account for Male-Killing Where Resource Reallocation Appears Unimportant?

What then of the other potential advantage to male-killing: a reduction in inbreeding and consequent fitness loss due to inbreeding depression? Unfortunately, we have precious little empirical data to allow assessment of the role such a mechanism has in the evolution and maintenance of male-killers. In only two cases is there direct evidence collected specifically to address this question. In one of these cases, the level of inbreeding depression in the wild was measured from egg clutches of *A. bipunctata* collected in Cambridge. Inbreeding was found to be rare in the wild and so unimportant with respect to male-killing in this population (Hurst et al. 1996). The other case, which concerned a recently discovered male-killer in the gypsy moth, *Lymantria dispar*, is discussed below.

For many of the species harboring male-killers, too little is known about the population ecology to allow assessment of the potential role of inbreeding avoidance in the spread of a male-killer. Of 19 cases known in 1993, this question could only be addressed in 11. Even in these, the assessments made were felt to be highly speculative, due to the lack of directly appropriate field data. In only two of these cases was it thought possible that inbreeding was likely to occur frequently enough in wild populations to be influential, although convincing evidence is lacking for both. One case involves the parasitoid wasp *N. vitripennis*. This wasp frequently inbreeds, for males emerging from their host fly puparia frequently mate with their sisters that emerge from the same puparium. However, although levels of inbreeding depression in this species are not known, it seems likely that they are probably low, since males are haploid and so cannot harbor deleterious recessives without them being expressed (p. 72). Inbreeding depression is thus confined to those genes whose expression is limited to the diploid females. The male-killer in *N. vitripennis*, a γ-proteobacterium, has a rather high vertical transmission efficiency, so the level of compensation needed to maintain the male-killer is quite modest and inbreeding avoidance might be sufficient here.

In *E. postvittana*, mating occurs before moths disperse from pupation sites and it is probable that inbreeding occurs quite commonly, particularly in small colonies. The extent of inbreeding depression has not been quantified, but there is no reason to suggest that it would be abnormally low. Thus, in this case at least, avoidance of inbreeding could be responsible, wholly or in part, for the spread of a male-killer.

A further case has been reported subsequently. This involves the gypsy moth, *L. dispar*. This moth has a reputation in the field of sex ratio studies, for it was the subject used by Goldschmidt in his extensive investigations of variations in sex-determining factors over geographic distance, during the first half of the 20th century (e.g., Goldschmidt 1934, 1940). From his work, Goldschmidt concluded that there were different geographic races of *L. dispar* with respect to sex determination, some races being categorized as "strong"; others "weak." Crosses between males and females of different races resulted in high production of sexual abnormalities or biased sex ratios. For example, crosses between a "strong" male race and a "weak" female race produced all-male families. Conversely, if female progeny from "strong" females crossed with "weak" males were then crossed with "weak" males of their father's race, only daughters were produced, some of these being normal females while others were feminized males (chromosomally ZZ but with female phenotype).

In a recent study, using gypsy moths from the Japanese island of Hokkaido (characterized by Goldschmidt as "very weak"), Yasutomo Higashiura and his colleagues describe the production of all-female sex ratios in a lineage being associated with approximately 50% egg-hatch rates (Higashiura et al. 1999). The sex ratio trait was shown to be maternally inherited with very high efficiency, 26 families from this single sex ratio line producing 964 females and no males. The trait was also shown not to be due to parthenogenesis, six nonmated females from the line producing only infertile eggs that failed to hatch. Although no antibiotic treatment was administered to the SR line and it was not examined microscopically or with molecular genetic tools, available data does suggest that a cytoplasmic male-killer is probably responsible. Interestingly, this male-killer appears to impose a reproductive cost on females that carry it, for these females produced significantly fewer eggs than other females. Assessment of the field prevalence of the SR trait showed it to be present in about 9% of females.

Selective factors that might be responsible for the existence of the male killer in this species were investigated. An advantage due to resource reallocation is dismissed because the larvae of this species disperses widely so that the reduction in competition from the lack of male larvae would not be directed toward infected females. Furthermore, Higashiura et al. reported that they never saw larvae feeding on unhatched eggs. In addition, a strong resource reallocation advantage seems unlikely, given that the larvae feed on the leaves of trees and strong competition specifically between siblings is likely to be low.

The level of inbreeding depression was also assessed in the laboratory by looking at the reduction in egg-hatch rates from brother/sister matings compared to matings between unrelated individuals. The level of inbreeding depression was extreme, with only 58% of eggs hatching compared to 95% from the outbred families. However, the level of inbreeding in the wild, obtained by determining the hatch rates of field-collected egg clutches was low. Higashuira et al. (1999) thus argue that avoidance of inbreeding cannot explain the presence of the male-killer in this species and conclude by saying that a new hypothesis may be needed to explain the evolutionary advantage of male-killing in *L. dispar*. It would be helpful if the agent responsible for the SR trait were identified and more extensive data on both inbreeding rates (e.g., after a population crash) and sibling interactions were collected in this case. However, as we will see, other recently discovered instances of male-killers in the Lepidoptera also appear difficult to attribute entirely to the classical advantages accruing from male-killing.

The Extraordinary Case of the Nymphalid Butterflies

There is one set of male-killers that have recently been found, which I have specifically not mentioned until now. These are the male-killers that have been discovered in four African nymphalid butterflies from two quite different genera. Three of the hosts belong to the genus *Acraea*. The fourth species is the widespread tropical/subtropical butterfly, *Danaus chrysippus*. The male-killers in these two genera pose some very interesting problems and reveal some truly amazing features. Because the problems posed in the understanding of male-killing in the two genera are rather different, I will deal with them separately, starting with the *Acraeas*.

Biased Sex Ratios in Acraea *Butterflies*

The butterfly *Acraea encedon* (Plates 7a–c) has long been known to produce biased sex ratios. Poulton reported, as early as 1914, that some females only had female progeny. Populations in which some females produce only daughters occur widely across Africa, with the proportion of females showing the SR trait varying between countries. This sex ratio trait has been shown to be inherited maternally (Owen 1970). In the Lepidoptera, females are the heterogametic sex. Thus, maternally inherited traits may be a consequence of genes on the W chromosome, as well as those on mitochondria or in inherited endosymbionts. Indeed, the all-female broods in *A. encedon* were initially ascribed to a driving W chromosome that distorted the primary sex ratio. Chanter and Owen (1972) reached this conclusion because they observed that the number of progeny produced in laboratory all-female broods was similar to the number produced in normal broods. They thus reasoned that a male-killer could not be involved, since this would lead to a reduction in the number of progeny in all-female families. However, given that survival of laboratory-bred butterfly larvae is frequently density dependent, this reasoning was potentially flawed and the case was re-examined.

More Hosts of Male-Killing Wolbachia

Initial work was on samples of *A. encedon* collected in 1997 on the Ugandan shores of Lake Victoria. Egg-hatch rates and offspring sex ratios were obtained from 35 females. Of these only 5 produced both sons and daughters, the other 30 all producing just daughters (Jiggins et al. 1998). In none of the 30 all-female families were the egg-hatch rates over 50%, while in the normal families hatch rates up to 100% were recorded, leading to the suspicion that the trait was due to a male-killer. This was confirmed by feeding larvae from all-female broods leaves dipped in antibiotics (tetracycline or sulphamethoxazole), whereupon the resulting adults produced both male and female progeny. Molecular genetic analysis then revealed the bacterium in question to be a *Wolbachia* (Hurst et al. 1999).

The finding of a second male-killing *Wolbachia*, in addition to that already described from *A. bipunctata*, is significant for three reasons. First, it shows that male-killing *Wolbachia* are not restricted to a single group, the coccinellids, which, due to their specific ecology, are particularly prone to male-killers. Second, it raises a question about estimates of the proportion of insect species that are infected by *Wolbachia* obtained from surveys involving a small number of individuals of a potential host species (e.g., Werren et al. 1995a). The prevalence of male-killers is typically in the range of 1–50%, although there are exceptions. This being the case, estimates of the distribution of *Wolbachia* or other ultraselfish bacteria based on assaying just a few individuals of a species are bound to be underestimates. Third, and perhaps most important, it shows that not all male-killing endosymbionts detect that they are in males by identifying a region of the Y chromosome, for while males are XY in *A. bipunctata*, it is the females that are heterogametic in *A. encedon*.

Exceptionally Female-Biased Populations

The proportion of females in the 1997 sample of *A. encedon* that produced just daughters (86%) was very high compared to other male-killers. However, this level was in line with material previously published by Owen and Chanter (1969), although their data has to be treated with some caution, since they included data from populations that subsequently were shown to belong to two different species, *Acraea encedon* and its sister species, *Acraea encedana*.

In 1998, samples of both *A. encedon* and *A. encedana* were collected from a variety of sites in southern Uganda, particularly around Kampala. The *Wolbachia* male-killer was found to occur in every one of these samples from both species and, although the prevalences varied, in many samples the prevalence was as high as, or higher than, in the original sample (Jiggins et al. 1999b, 2000b). The male-killer in *A. encedana* is the same as that in Ugandan *A. encedon*: a male-killing *Wolbachia* that is maternally inherited, is antibiotic sensitive, and kills male embryos before they hatch from eggs. This *Wolbachia* has relatively high, but not perfect, vertical transmission. Due to the high prevalence of the male-killer in both these species, many populations exhibit extremely female-biased sex ratios. For example, at one Ugandan site, only 7 males were found among 1,475 *A. encedon.* At another site, 52 males were found in a sample of 871 *A. encedana.* Interestingly, a second strain of male-killing *Wolbachia* was found in Tanzania, where the population was polymorphic for both *Wolbachia* strains. This is similar to the situation in some Russian populations of *A. bipunctata.*

I will leave the evolutionary problems and consequences presented by populations in which over 95% of butterflies are female to the next chapter (p. 168). Here I will focus on the problem of how the male-killer may have reached this very high prevalence. By collecting larval nests from the wild and testing them for the presence of *Wolbachia*, the vertical transmission efficiency of the *Wolbachia* in *A. encedana* was estimated to be 0.96 (Jiggins et al. 2000a). On the basis of this estimate and assuming that the male-killer was at equilibrium, the lifetime reproductive output of infected relative to uninfected females was estimated to be at least half as great again (minimum estimate within 95% confidence limits = 1.55). This is extraordinarily high. A similar estimate made for English *Rickettsia*-bearing *A. bipunctata* was only 1.16 times that of uninfecteds.

The problem in these *Acraea* butterflies compared to coccinellids is that it is not evident how infected females benefit from the death of males. The eggs of both species are laid in large batches of 100+. Neonate larvae consume their eggshells and some, but not all, unhatched eggs. The larvae from a clutch remain together after they disperse from their egg mass and feed aggregatively until splitting up in the final, fifth instar (Plate 5e). These features make resource reallocation a possible source of fitness compensation, both through sibling egg consumption and through reduced competition for food. However, in neither species is there any evidence that larval foodplant is strongly limiting and it is difficult to conceive that resource reallocation alone could be responsible for the high prevalence level of the *Wolbachia* in light of the incomplete vertical transmission. Another possible explanation for the high estimated value that is required to explain the prevalence data is that the assumption that the population is at equilibrium is not valid. This possibility must be weighed against data going back through much of the 20th century suggesting that in some populations of *A. encedon*, and probably *A. encedana*, strongly female-biased sex ratios have existed over hundreds of generations (e.g., Owen and Chanter 1969; Gordon 1984; Owen et al. 1994).

Another possible explanation is highlighted by recent work by Emily Dyson on another nymphalid butterfly, *Hypolimnas bolina*, in Fiji. *Hypolimnas bolina* has long been known to produce all-female families (Simmonds 1926). These families were shown to involve the death of males (Clarke et al. 1975) and Dyson has shown the trait to be associated with the presence of a *Wolbachia* (Dyson 2000). The *Wolbachia* occurs at high prevalence and has a high vertical transmission efficiency. Perhaps most interestingly, Dyson found that the mortality of uninfected larvae was significantly greater than that of infected female larvae reared under the same conditions and at the same density.

This raises the question of whether the high prevalence of male-killers in *A. encedon* and *A. encedana* result from a direct benefit that they confer on female bearers. However, the answer to this question is *no*, because experiments to test whether *Wolbachia* confer any benefit on female bearers show that in *A. encedon* they do not.

The reason for the high prevalence of male-killing *Wolbachia* in *A. encedon* and *A. encedana* remains unresolved. Of course, as-yet-unrecognized factors may be of relevance in the dynamics of male-killers and their hosts. What these might be is open to imaginative speculation. For my part, I offer one suggestion in chapter 7, where I note the correlation between male-killer presence and color pattern polymorphism in some insects with true warning colors (p. 176).

Wolbachias *in Other* Acraea *Species*

Before leaving the *Acraea* genus, mention should be made of a third *Wolbachia*-bearing species in which female-biased sex ratios have been observed and of several other species that have been shown to host *Wolbachias*, but in which their effect is either some manipulation other than male-killing or is unknown.

In 1998, samples of a variety of species of *Acraea* butterfly were made in Uganda, to allow these to be assayed for *Wolbachia*. From these collections, eight more species of *Acraea*, in addition to *A. encedon* and *A. encedana*, were found to be infected. These eight can be split into three groups: those in which all butterflies were found to have *Wolbachia*; those in which *Wolbachia* were found in both males and females but some butterflies were free of infection, and those in which only females were infected (see Table 6.9).

Considering these categories separately, the first group may represent cytoplasmic incompatibility–inducing *Wolbachia*. The high prevalence of *Wolbachia* in these species and the presence of the bacterium in both males and females are characteristic of CI infection. Indeed, in *A. acerata*, data from breeding experiments using infected and antibiotic-cured butterflies shows that the *Wolbachia* in this species does cause CI.

In the second group, the low level of uninfected individuals in *Acraea pentopolis* and *Acraea pharsallus* may mean that these are also CI inducers, but sample sizes are rather low, particularly in the former. In *Acraea equitoria*, again the sample size is small, but here CI seems to be a less attractive possibility, since CI *Wolbachia* are usually at frequencies close to fixation. The third group are perhaps the most interesting, for the presence of *Wolbachia* just in females strongly suggests that the *Wolbachia* is expressed as a sex ratio distorter.

Studies of one of these, *Acraea eponina* (Plate 5a), confirmed this suggestion, when a number of field-collected larval nests (larvae of *A. eponina* remain in family groups) produced just female offspring. Breeding data, antibiotic

TABLE 6.9
Species of *Acraea* shown to harbor *Wolbachia*. The number of males and females that tested positive (+) and negative (−) for *Wolbachia* are given.
(Data from Jiggins 2000.)

	Males		*Females*		
Species	+	−	+	−	*Country*
A. acerata	20	0	51	0	Uganda
A. alcinoe	4	0	22	0	Uganda
A althoffi	11	0	12	0	Uganda
A. eponina	—	—	5	15	Uganda
A. equitoria	2	4	0	3	Colombia
A. macarista	0	15	3	11	Uganda
A. pentapolis	7	0	2	1	Uganda
A. pharsalus	5	1	21	0	Uganda

treatment, and molecular analysis demonstrated that the SR trait is due to the *Wolbachia*. However, although the egg-hatch rates in SR broods were lower than those in N broods, the mean hatch rates of the SR broods were greater than 50%, and in many SR families over 70% of eggs hatched. This casts doubt over whether the SR *Wolbachia* in this instance is acting as an early male-killer. It could, of course, be acting as a late male-killer, but there is no evidence of horizontal transmission of bacteria from corpses favoring delayed pathogenicity in this genus. Another possibility is that the *Wolbachia* is causing a distortion in the primary sex ratio (p. 117). Further work on this species is likely to be of great interest.

The Case of Danaus chrysippus*: A Male-Killer with No Obvious Advantage*

Danaus chrysippus (Plates 7e and 7f) is a widely distributed nymphalid butterfly of the Old World tropics. In East Africa, this butterfly exhibits two unusual features. First, the color patterns of this warningly colored and chemically defended butterfly are polymorphic. Elsewhere, the species is monomorphic. Second, here, and, as far as is known, only here, a high proportion of females exclusively produce daughters as a result of the death of their sons as embryos or young larvae (Smith et al. 1998). The SR trait is maternally inherited, with only rare loss of the trait (Owen and Chanter 1968; Smith et al. 1998). This pattern is thus suggestive of another bacterial male-killer. However, there is a problem. *Danaus chrysippus* females lay their eggs singly on widely scattered larval foodplants (various milkweeds) and larvae feed singly (Plate 5f). Smith et al. argue that, based on this oviposition behavior, it seems unlikely that siblings interact to any appreciable extent; as a result, the advantages of resource reallocation necessary for the spread of a male-killer seem to be lacking here. The dispersal characteristics of *D. chrysippus* also make inbreeding avoidance an unlikely source of fitness compensation. Instead, Smith et al. (1998) contend that the East African region is a hybrid zone between races of *D. chrysippus* and that both the color pattern polymorphism and the female-biased sex ratios are a consequence of hybrid matings. In the case of the color patterns, different sets of color pattern genes from the different races are mixed in hybrids, and these then segregate in subsequent generations. The female-biased sex ratios are explained to be a consequence of a mitochondrial gene, which, while benign on the genetic background of its own race, causes the death of males on a hybrid genetic background.

As with the meiotic drive explanation of all-female broods in *Acraea* butterflies, this hypothesis did not appear totally convincing. Consequently, collections of *D. chrysippus* were made in East Africa in 1997 and 1998. Approximately one-third of the females produced female-biased sex ratios (Jiggins et al. 2000b). The egg-hatch rates of these families were considerably lower than those of clutches from females that produced normal sex ratios. In addition, the death rate in first instar larvae from SR females was lower than from N females. Putting these two together, the proportion of offspring that reached the second larval instar from SR mothers was almost precisely half (41%) that of N mothers (83%). When antibiotic treatment cured some SR lineages of the trait, a male-killing bacterium seemed the likely cause of the sex ratio bias. Probing infected females with DNA sequences specific to a range of male-killers showed a *Spiroplasma* to be present in SR but not in N lines. Phylogenetic analysis showed that this *Spiroplasma* was closely related to the male-killing *Spiroplasmas* of the ladybirds *A. bipunctata* and *H. axyridis* and to a symbiont of a tick, *Spiroplasma ixodetis*. Limited searches for the *Spiroplasma* in samples from outside of East Africa failed to find it.

These results lead to the conclusion that the cause of male-killing in this species is not mitochondrial, as suggested by Smith et al. (1998), but rather is a bacterium. Of course, this does not preclude male-killing involving a hybrid zone. If the *Spiroplasma* had coevolved as a benign symbiont of one particular race, then it might become spiteful on a hybrid genetic background. This type of pattern is seen in a species of fruit fly, *Drosophila paulistorum*, in which male sterility in hybrids is caused by an inherited bacterium that on its parental genetic background is thought to be beneficial to its host (Ehrman and Kernaghan 1971). However, were this to be the case in *D. chrysippus*, expectation would be that the *Spiroplasma* would be present at very high frequency in one host race outside of East Africa, and this does not appear to be the case.

So there is a problem. What advantage could there be to male-killing in the East African population of this species? The answer is as yet unknown. It is possible that simply because *D. chrysippus* usually lays eggs singly, it is premature to discard resource reallocation as conferring some advantage. Certainly, eggs are occasionally laid in twos or threes, and I have observed larval cannibalism of both this species and *Danaus plexippus* in Australia. Indeed, Ackery and Vane-Wright (1984) report larval cannibalism in both of these species. Yet, given the imperfect vertical transmission of this *Spiroplasma*, and the migratory behavior of its host, it is difficult to conceive of an advantage through resource reallocation sufficient to maintain the high prevalence (up to 40%) of the bacterium in host populations in Kenya and Uganda. Other possible factors now need to be investigated. Horizontal transmission is feasible, although on the basis of Laurence Hurst's theories, the expectation would be that male-killing would occur late in larval life. Inbreeding avoidance seems unlikely because the species is highly mobile and males cannot mate for several days after eclosion (Schneider 1987). Maybe here, as with *L. dispar*, we have an instance where we need to look for a novel advantage to male-killing.

There are a number of intriguing questions that the male-killing *Spiroplasma* of *D. chrysippus* raises. First, male-killing *Spiroplasmas* are of two distinct types, those of group II, which cause male-killing in some *Drosophila* species, and those of group VI, which cause male-killing in *D. chrysippus* and two ladybird species. With respect to the latter, do these represent two independent evolutions of male-killing, or has there been horizontal transmission between the hosts? If male-killing has only evolved once in this group of *Spiroplasmas*, it raises the interesting possibility that butterflies and beetles, which have different sex chromosomes (beetles are XY male heterogametic; butterflies are ZW female heterogametic), have a common feature in their male determination systems that can be identified by the *Spiroplasma*. We may also ask whether these *Spiroplasmas* are male-killer specialists, or whether, like *Wolbachia*, they have flexibility in the manners with which they manipulate their hosts.

Finally, there is one further extraordinary feature of the *D. chrysippus* case that may be mentioned. The SR trait in *D. chrysippus* was found by Smith et al. (1998) to be associated with particular color patterns. Not only that, but these color patterns were also controlled differently in male and female butterflies, being sex linked in females and either sex linked or autosomal in males. Is this link coincidental, or is there some adaptive link between male-killers and color patterns in some species? I will return to this question in chapter 7 (p. 176).

The Mechanism of Male-Killing

Very little is known about the mechanism of male-killing. Few investigations into the mechanism of early male-killing have been undertaken and these are urgently needed. Such research needs to answer two questions. First, how does the male-killer determine that it is in a male rather than a female embryo? Second, how does it then kill the males?

Perhaps the best evidence pertaining to the way male-killers determine the sex of their host comes from work on *D. melanogaster*. Here, specialized mutant stocks with abnormal chromosome complements (e.g., triploid, or with the sex chromosomes attached to autosomes) were artificially infected with male-killing *Spiroplasma* from other *Drosophila* species (Sakaguchi and Poulson 1963). These experiments showed that the Y chromosome is not involved in male determination. Rather, only those individuals with two X chromosomes survived the infection. Single-X individuals died, suggesting that the male-killer

TABLE 6.10
Examples of some types of male-killing bacteria and the different sex determination systems of their hosts.

Wolbachia	*Spiroplasma*	γ-*Proteobacteria*
Male heterogametic	Male heterogametic	Male heterogametic
e.g., *A. bipunctata*	e.g., *H. axyridis*	e.g., *C. sexmaculatus*
Male homogametic	Male homogametic	Male haploid
e.g., *A. encedon*	e.g., *D. chrysippus*	e.g., *N. vitripennis*

is capable of counting X chromosomes, or quantifying some product produced only when two X chromosomes are present.

It is far from certain that the same mechanism of male-determination can be used by all male-killing bacteria. This is because the sex determination mechanisms of hosts are very different (Table 6.10). For example, in *Drosophila*, ladybirds, and milkweed bugs, males are heterogametic, while in the Lepidoptera, males are homogametic. It is perhaps worth noting here that, in most male-killer hosts, no work has been conducted to establish how sex is determined beyond examining them to see which is the heterogametic sex. Thus, for example, although there are more species of beetle in the world than any other type of organism, the mechanism of sex determination has not been demonstrated in a single species.

Critical to investigations of the mechanism of male-killing is the precise timing of male death. In *Drosophila* and *Rickettsia*-infected *A. bipunctata*, male death appears to occur very early in embryogenesis (e.g., Counce and Poulson 1962). In other cases, such as the *Acraeas* and ladybirds infected with *Spiroplasma*, death is often later in embryogenesis, so that the developing embryo can be seen through the eggshell. There is some evidence that in *H. axyridis*, some infected male embryos hatch and only die in the first larval instar. The same is true for *D. chrysippus*, and this may be a feature of male-killing by this type of *Spiroplasma*. Observations on a variety of species have suggested that the timing of male death may be influenced by bacterial density. Indeed, work on *A. decempunctata* has indicated that if the density of male-killing *Rickettsia* in males of this species is very low, some may survive to maturity. It is to be hoped that the finding of a gene in one species of ladybird that prevents the killing action of a male-killer (p. 180) may help in the search for the mechanism of male-killing in a species in which the costs and benefits of male-killing are also known.

Sperm-Killers and Distortions of the Primary Sex Ratio

The timing at which male-killers kill their male hosts is an important aspect of their evolutionary strategy. It is therefore worth considering how early in reproduction male-killing could occur. Could a symbiont in an infected female of a female homogametic species kill Y-bearing sperm after copulation but before fertilization? In many insects this must be an option, since sperm is stored, often for substantial periods, in the female's spermatheca. In this event, a female-biased sex ratio would be produced, but the egg-hatch rate would not be abnormally low. If such a case did exist, it would produce a bias in the primary sex ratio, but would be distinct from an X chromosome meiotic driver because the ratio of X- to Y-bearing sperm produced and passed by males to females would be normal.

In female heterogametic species, symbiont-induced death of W-bearing female gametes would also produce a bias in the primary sex ratio, and this would be a case of microbe-induced meiotic drive (p. 76).

Finally, it is possible that a symbiont may produce a female-biased sex ratio by killing males after fertilization but before egg laying or birth. In most species that are known to harbor male-killers, fertilization of fully resourced female gametes occurs just before oviposition. In these species, there appears little scope for symbionts to have an effect in the short time between fertilization and the eggs being laid. However, in species in which the embryo is resourced within the mother after fertilization, it is possible that there could be circumstances when killing males prior to birth could be advantageous. For example, if the resources that would have gone to male embryos were diverted to female embryos, the male-killer would gain an advantage and spread. Here, again, the distortion would appear to be in the primary sex ratio, even though male-killing was responsible. Cases of this type are likely to be rare and very difficult to detect, if they exist at all. However, they would certainly be worth seeking.

Having described the basic strategy of male-killers, their diversity, and the way in which they gain advantage by killing male hosts, it is worth considering the effect that these bacteria have on the evolution and behavior of their hosts. It is to the evolutionary consequences of male-killing that I now turn.

CHAPTER 7

The Evolutionary Consequences of Sex Ratio Distortion

> When you have eliminated the impossible, whatever remains, however improbable must be the truth.
>
> —Sir Arthur Conan Doyle, *A Study in Scarlet*

Summary

The aim of this chapter is to consider the ways that sex ratio distortion due to inherited symbionts may impact on the evolution and genetics of their hosts and how evolutionary responses of the host may influence the distorters' evolution and behaviors. A variety of features, from clutch size and sexually transmitted diseases to reproductive strategies and mitochondrial DNA variability are considered. In addition, host responses to the sex ratio biasing action of inherited symbionts, through increased production of males or through the evolution of suppressors are discussed.

Empirical evidence relating to the evolutionary impact of symbiont-induced sex ratio distortion is scarce. However, some data are already showing that these microorganisms may have a profound impact on the evolution of their hosts. The theme of this chapter is that many ecological and behavioral characteristics of host species are only likely to be comprehensible if the role of sex ratio distorters is understood against a context of these traits.

Evolutionary Oddities?

From an analysis of the literature, one might get the impression that male-killers are one of the least important of the sex ratio distorters. Yet a consideration of the range of taxa that have been found to harbor male-killers and the taxonomic diversity of the micro-organisms concerned argue against this suggestion. The paucity of literature concerning male-killers prior to the 1990s is the result of several factors. Sex ratio biases caused by male-killers have been attributed incorrectly to other causes (e.g., sex chromosomal meiotic drive, hybrid zones). Male-killers are difficult to detect because most have relatively low prevalences and vertical transmission rates compared to other inherited symbionts. The causative agents of male-killing were difficult to establish prior to the development of appropriate molecular genetic methodologies for the identification of microorganisms. Finally, relatively few scientists have turned their attention to this phenomenon.

I have no doubt that cases of male-killing will be found in other insect orders, in addition to the five in which they have thus far been reported. I also suspect that it is only a matter of time before they are reported far more widely in other arthropod taxa, particularly the arachnids, many of which have behavioral and ecological traits that should make them prone to male-killer invasion. Nor is it likely that male-killing will prove to be confined to the arthropods. Some groups of invertebrates outside the Arthropoda have attributes that are likely to make them susceptible.

The rate at which new examples of male-killers have been found has increased rapidly over the last ten years and is accelerating. The view that we have of these sex ratio distorters is changing from that of evolutionary oddities, mere curios that can safely be left as a footnote in our considerations of the history of life on Earth, to a more central position. It is becoming clear that these inherited microbes have a considerable influence on many aspects of the biologies of the hosts they inhabit.

Male-Killers and the Evolution of Clutch Size

I will begin with a behavioral trait that at first sight appears to be simple—the number of eggs that a female lays together in a batch. From work on birds, there is strong evidence that animals produce clutch sizes that maximize their own fitness. The number of eggs laid is a tradeoff between the number of eggs that can be produced and the parents' ability to successfully feed and protect the young. In terms of insects harboring male-killers where parental care is rarely significant, we can think of the tradeoff as being between the cost to the mother, in time and energy, of finding suitable sites to lay her eggs and the costs to her offspring in terms of sibling competition. One of the most obvious impacts of early male-killers is that in a clutch of eggs from an infected female, the proportion that hatch is lower than in a clutch of the same size from an uninfected female. The level of competition between the larvae that results from such a clutch, therefore, will be lower than between larvae from a normal clutch. Consequently, we may expect that females can afford to lay larger clutches of eggs, thereby reducing the number of oviposition sites that must be sought (Hurst and McVean 1998).

An increase in the size of egg clutches in male-killer–bearing individuals has been sought in coccinellids. However, the predicted increase in clutch size has not been found. Comparisons of clutch sizes from populations with different male-killer prevalences either show no differences in average clutch size or produce a negative correlation of clutch size with male-killer prevalence. There are several possible reasons for the lack of the response predicted by Hurst and McVean.

First, unless a female can detect that she is bearing a male-killer, the increase in clutch size resulting from male-killing will be spread across all members of the population, irrespective of whether they are infected or not. This will mean that uninfecteds lay sub-optimal clutches. For clutch size to be affected, the prevalence of the male-killer would have to be high.

Second, the cost of bearing a male-killer may impact on female nutrient resources and so limit the number of eggs that a female can mature at one time for laying in a clutch. This accounts for the fact that in some of the species studied, infected females laid, on average, smaller clutches than uninfected females.

Third, other factors may limit changes in clutch size. For example, the small Japanese ladybird, *Coccinula sinensis*, which harbors a male-killing Flavobacterium, lays exceptionally large eggs for an aphid-feeding coccinellid when assessed in relation to the weight of its mother (Majerus and Majerus 2000a). An average egg of *C. sinensis* is 1.75% of the weight of the mother, compared to figures of 1.09% and 0.57% for *Propylea japonica* and *Cheilomenes sexmaculatus*, respectively. This has been attributed to there being a minimal size for neonate larvae at dispersal if they are to have a reasonable chance of catching an aphid. The level of nutrients contributed by a female *C. sinensis* to each of her large eggs, means that she lays a rather small number of eggs in each clutch (average = 5.57 eggs; range = 1–14 eggs). In all other ladybirds in which male-killing has been reported, average clutch size is greater than ten. Consequently, selection for increased clutch size, resulting from male-killing as suggested by Hurst and McVean's hypothesis, is likely to have very little effect in *C. sinensis*. In this species, it seems that clutch size is limited by the level of nutrients available to resource eggs, which, in turn, results from the need for hatching larvae to attain a certain size. Assessments of clutch sizes, both in the laboratory and in the field, have shown no appreciable difference between clutch sizes laid by infected and uninfected females. Such constraints are not likely to apply to all species that harbor male-killers and work to test the impact of male-killers on clutch size would certainly be worthwhile.

The Sex Ratio and Sexual Selection

As we have already seen, the sex ratio may have a considerable effect on the courtship and copulatory behavior of animals. When males are common, they tend to compete with one another for access to females and females can afford to be, and indeed should be, choosy. In such populations, selection will favor males that are strong, and can solicit, impress, and excite females with extravagant courtship displays or by providing nuptial gifts. High-quality males can also protect their paternity by high sperm production or by preventing females from remating by guarding the female directly or by insertion of sperm plugs or antiaphrodisiac chemicals. This is the pattern in most species in which the sex ratio of available, reproductively mature adults is close to parity or is male biased. However, in populations that are female biased because of the presence of male-killers, the strength of selection for males to compete and females to choose is reduced. Generally, the greater the female bias, the greater this reduction, although the mating system, particularly the potential levels of promiscuity of each sex, may mediate this reduction. If the sex ratio distorts very strongly toward females, such that the number of males rather than the number of females limits the population's reproductive rate, full sexual role reversal will result, with females competing with one another for copulation opportunities and males becoming choosy.

Male Mate Choice

Sex role reversal has been reported in a small but disparate array of species in which males contribute more to their progeny than do females. Thus, in the lily-pad-walking American jacanas, *Jacana spinosa*, females compete for territories. The males make and tend nests, care for the offspring and choose between competing females on the basis of the resources contained in each female's territory.

In the dance fly, *Empis borealis*, males provide females with a nuptial gift of food before mating. This gift provides the female with much of the nutrition used in forming her eggs. The male is thus contributing greater resources for the offspring than the female. This is reflected in the behavior of the dance flies. Many species of animals exhibit what is called *lekking behavior*, where males congregate at a particular site (usually lacking in resources), there to be visited by females willing to mate, who choose a partner from the assembled males. Obvious vertebrate examples include Ugandan cob, fallow deer, sage grouse, ruffs, and African Talapi fish. Among insects, male leks have been reported from many orders, including the Diptera and Lepidoptera. However, unusually, in the dance flies it is the females that form lekking swarms that are visited by males, who then choose between the females, their choice being for the largest females available (Sivinski and Petersson 1997).

The sex ratio distortions produced by male-killers provide elegant testing grounds for theories of the evolution of reproductive strategies and mating behavior. In this context, it is perhaps fortunate that male-killers occur in a variety of different insect orders and because of differences in their vertical transmission efficiencies and prevalences, both within and between species, a range of female-biased sex ratios are available for investigation. A number of recent studies provide evidence that male-killers do indeed have highly significant effects on mating behavior and other aspects of host reproduction.

As we have already seen, the Asian ladybird *Harmonia axyridis* is infected by a male-killing *Spiroplasma*, which varies in its prevalence throughout Asia. The sex ratio of various populations is inversely correlated to the prevalence of the male-killer (Figure 6.7, p. 148). This species of ladybird has previously been shown to have rather flexible mating behavior. Painstaking experiments by Naoya Osawa and Takayoshi Nishida (1992) have shown that females collected in Central Honshu choose between males of different color patterns and that these mating preferences vary both between populations and within populations through time. More intriguingly, their work showed that there is an element of male choice in the system.

The sex ratios of both *H. axyridis* and *Adalia bipunctata* vary geographically

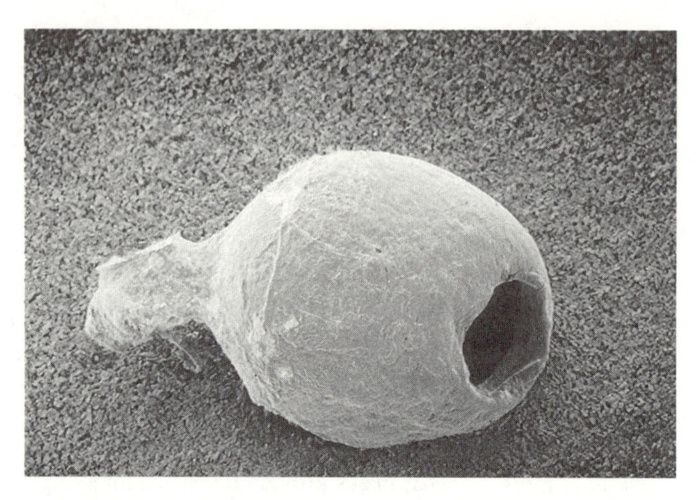

Figure 7.1 Electron micrograph of an empty and ejected spermatophore of the two-spot ladybird, *Adalia bipunctata*. (Courtesy of Mark Ransford.)

(p. 148), and mating preferences and copulatory behavior are well documented for both (see Majerus 1998 and refs. therein). These species thus provide perfect material to test whether the strength of mate competition, mate choice, and reproductive investment are correlated to the sex ratio, as theory predicts.

Male Reproductive Investment and the Sex Ratio

One possible effect of a female bias in the tertiary sex ratio is that males may invest less in each mating in severely female-biased populations than in populations in which the sexes are closer to parity. In *A. bipunctata* this has been shown to be the case.

Male *A. bipunctata* form spermatophores within their partners. Unusually, within a single copulation a male may insert one, two, or even three spermatophores. The female later ejects the spermatophore husks (Figure 7.1), so the number passed can be counted. A male two-spot ladybird may reduce his investment into each copulation either by reducing the number of sperm in each spermatophore, or by inserting fewer spermatophores per copulation.

Tests on three populations with different sex ratios showed that the number of sperm in each spermatophore did not vary between populations. However, the number of spermatophores passed did, with males from the more highly female-biased population passing significantly fewer spermatophores per copulation (Figure 7.2).

In the case of these ladybirds, males are relatively long-lived (up to a year), and can mate every day throughout the breeding season. Given the reported levels of sex ratio distortion in natural populations of ladybirds, it seems unlikely that females will die without mating simply as a result of the lack of

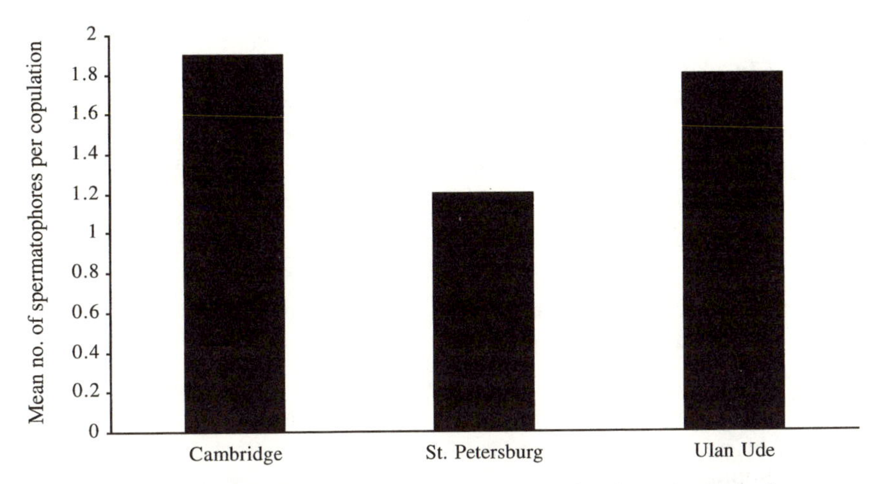

Figure 7.2 Number of spermatophores passed per copulation by males from three populations of the two-spot ladybird. The prevalences of male-killing bacteria in these populations were: Cambridge—0.06–0.11; St. Petersburg—0.5; Ulan Ude—0.03.

males. This is not the case for the butterfly, *Acraea encedon*, which harbors male-killing *Wolbachia*.

Sex Role Reversal and Female Leks

I first saw *A. encedon* in the wild in Uganda in 1997. Within a few hours of landing in Entebbe, I was sitting at the top of a hill called Tank Hill on the Campus of Makerere University, looking down over a grass-covered, western-facing slope in the early afternoon. Hundreds of butterflies were flying there, landing on the grass stems in little clusters and taking off again (Plates 5b–d). And as far as I could tell, all the butterflies were female. Indeed, on subsequent days, Francis Jiggins, Thomas Kyokuhaire, and I netted over 700 females at this site before a male was caught. It struck me that the swarms of females bore striking parallels with the male leks seen in many animals. The aggregation was on a resource-free site, the individuals interacted strongly with one another, and they were all of one sex. The only problem was that they were all female rather than all male. Could this be a case of full sex-role reversal, with males choosing between competing females?

Subsequent field and experimental work strongly suggests that it is. Populations of *A. encedon*, in southern Uganda, were caught, sexed, and either marked (to prevent recounting and to later assess longevity and dispersal) and released, or were retained to allow determination of male-killer infection. The sites were characterized as those with resources in the form of the larval foodplant, *Commelina*, and those without (Box 7.1a). The results (Box 7.1b) showed that the samples at resource-lacking sites, categorized as lekking sites, were both more female biased and showed higher male-killer prevalence levels than those from

Box 7.1a Comparison of the vegetative and topographical features of lekking and foodplant sites for *Acraea encedon*, with notes on the sex ratios, behaviors and male-killer prevalence of the butterflies observed at each site type (adapted from Jiggins et al. 1999b, with additional observations from Majerus unpubl.).

Lekking sites	*Foodplant sites*
Short to mid-length (up to about 70cm) vegetation, in close proximity to trees.	Variable general vegetation.
Most frequently, but not exclusively on west facing slopes near or on hilltops.	No particular topographical features evident, other than soil conditions favorable for *Commelina benghalensis*.
No larval foodplant or nectaring plants available for adult feeding.	Larval foodplant, *Commelina benghalensis* always common. Necturing plants sometimes present. Adults observed feeding at some sites.
Acraea encedon present predominantly after noon.	*Acraea encedon* from sunup to sunset.
Females fly within a small arena. If other females are encountered, they engage in brief midair chases and often land on the ground. Aggregations of females form on long grass stalks or other plants.	No abnormal female behaviors observed. Mating, adult feeding, oviposition all observed and apparently unexceptional.
0–12% butterflies male	34–50% butterflies male
Males rarely observed but a high proportion of those seen were *in copula*.	Males commonly observed. Mating pairs common, but proportion of males *in copula* lower than on lekking sites.
Prevalence of male-killer in females generally higher than at foodplant sites.	Prevalence of male-killer in females generally lower than at lekking sites.

Box 7.1b Details of samples of *Acraea encedon*, giving site location, site type, and the behavior, male-killer prevalence, and sex ratios of the samples. At sites where samples were taken on different dates, the data have been summed when neither the behavior of the butterflies nor their sex ratios differed between dates. (Adapted from Jiggins et al. 1999b.)

Location	*Behavior*	*Habitat*	*Male-killer prevalence (no. females tested)*	*Percentage male (sample size)*
Kagolomolo	Nonlekking	*Commelina* in farmland	0.81 (70)	34 (169)
Kibuyi	Nonlekking	*Commelina* in farmland	0.78 (27)	50 (56)
Gangu	Lekking	Lawn by tall trees on hilltop	0.87 (23)	0 (25)
Kawanda	Lekking	Grassy hilltop with small trees	0.70 (20)	0 (19)
Makerere	Lekking	Lawn beside trees on hilltop and slope	0.93 (215)	1 (2139)
Lubya	Lekking	Lawn beside trees on hilltop	0.92 (38)	2 (525)
Lake Mburu	Lekking	Lawn beside trees on ridge	0.97 (30)	6 (34)
Kajansii Hill	Lekking	Short grass in forest clearing on hilltop	1.00 (9)	0 (19)
Kajansii Swamp	Lekking	Grass pasture with *Eucalyptus* trees by swamp	0.92 (51)	12 (308)
Nalugala A (14/4/1998)	Nonlekking	*Commelina* in farmland, 50m from lekking site described below	0.86 (7)	47 (15)
Nalugala B (4/7/1998)	Lekking	Lawn beside tall trees by swamp	0.89 (9)	0 (10)
Nalugala B (7/27/1998)	Lekking	"	0.89 (9)	4 (24)
Entebbe A (4/7–17/1998)	Nonlekking	*Commelina* beside Lake Victoria, 75m from lekking site	0.79 (47)	46 (84)
Entebbe B (3/22/1998)	Lekking	Short vegetation surrounded by tall trees	1.00 (50)	2 (48)
Entebbe B (7/7–9/1998)	Lekking	"	0.96 (52)	2 (142)

foodplant sites. Jiggins then conducted a mark-release-recapture experiment to determine whether females were aggregating on the lekking sites specifically in order to mate. Females from families reared in captivity were marked and split into two equal groups. These females were then put into cages either with or without males. Those females that were observed mating were recorded. The following morning, the virgin and mated females were released at the lekking site on Tank Hill. Butterflies were then netted at this site in the afternoon of each of the next six days. The proportion of virgins that were recaptured (67 of 141 = 48%) was significantly higher than the proportion of nonvirgins (12 of 72 = 17%), showing that females visit the lekking sites in order to mate, or remain on the lekking sites for longer if they are virgin.

Can Males Choose?

A host with an efficiently inherited high prevalence male-killer could be in danger of extinction due to the lack of males. Such a fate might be averted if males could distinguish between male-killer–infected and –uninfected females and preferentially mate with the latter. In no case has it yet been demonstrated that males have this ability. However, further work on *A. encedon*, on Tank Hill, showed that uninfected females on the lek were significantly more likely to have mated than were infected females (Jiggins et al. 1999b).

Here then we have a small piece of evidence that males may be able to distinguish between infected and uninfected females. Certainly there is likely to be very strong selection in favor of this ability. Males that mate with uninfected females will produce sons. Because the sex ratio of *A. encedon* in this region averages about 95% female:5% male, sons will have a reproductive success almost 20 times that of daughters.

The deduction that males can preferentially choose to mate with uninfected females, must, however, be treated with some caution until the mechanism behind such choice is determined. It is feasible that the pattern of uninfected females more often having mated is the result of changes in the mating behavior of the species that do not depend directly on an ability of males to determine infection status of potential partners. For example, selection acting on both sexes may favor brother/sister matings either directly or indirectly. I will consider the situation for females first.

The Attractions of Incest

Acraea encedon lays eggs in very large batches (200–300). Larvae remain together until the final larval instars. It is thus likely that the siblings will be in fairly close proximity when they pupate and subsequently emerge. Furthermore, the members of a clutch emerge over a fairly short period of time. As mating can take place soon after the females have emerged, there seems to be a fairly high probability that at least some of the butterflies in the vicinity will be siblings. For SR females, this will not help for mating purposes, as all siblings will be female. However, for N females, half of these siblings will be male. Sib-sib mating is usually considered disadvantageous because of the inbreeding depression that can result. However, given the proportion of female *A. encedon* that die virgin and so have no reproductive success at all, it may pay for a female to mate with her brother rather than risk never finding a mate at all. Thus the extreme female bias in the sex ratio of this species will select for females that take any opportunity to mate that presents itself, even if this promotes mating with a male sibling close to their pupation site.

As a rider to this, one might question why any females who have mated should visit the female leks. One possibility follows from the argument that females should take the opportunity to mate whenever a male presents himself, even if he is a brother. Females that have mated close to their emergence site (and so may have mated with a sibling) will gain an advantage in mating again on a lekking site with an unrelated partner. Although highly speculative, I find this rationale persuasive. I would thus predict that the progeny from mated females caught as they arrive at a lekking site will, on average, be more homozygous (as a result of inbreeding), than progeny from mated females caught leaving the lekking site. An alternative to this scenario exists. If females are in excess, then just as already argued in the case of *A. bipunctata* (p. 170), males of *A. encedon* may gain an advantage by investing less sperm per copulation. If a female receives a small sperm load from her first male partner, she may become "sperm-exhausted" before she has laid all her eggs. She may therefore visit a lekking site to replenish supplies.

Considering males, the increased likelihood of mating with a sibling due to temporal and spatial proximity is not the only rationale behind the possibility of sib-sib mating in *A. encedon*. Even if siblings were distributed randomly within the population through both space and time, it might pay for a male to copulate with his female sibs, assuming that he has the ability to distinguish his sisters from other females. The logic is straightforward. If a male can identify female siblings, by mating with them he will almost certainly be mating with an uninfected female. His own survival to adulthood makes it highly likely that his mother and consequently his sisters do not carry the male-killer. This rationale for mating with sisters is thus the same as that favoring general choice of uninfected over infected partners, for a male's own sisters are one subset of the population that are likely to be free of the male-killer.

Whatever the reason that uninfected females have a mating advantage over infected ones, this system may be crucial to the long-term survival of this species and its sister species, *Acraea encedana*, which parallels *A. encedon* with respect to most features of male-killing. The increased likelihood of mating that uninfected females enjoy will act to prevent the complete eradication of males of the species by the spread of the male-killer, thus preventing extinction.

Another case of a mating advantage of uninfected compared to infected females involves *Wolbachia*-induced feminization. Models of feminizing *Wol-*

bachia in *Armadillium vulgare* predict that the prevalence of the *Wolbachia* ought to be higher than is observed in practice. Jerome Moreau and his colleagues at the University of Poitiers tested whether uninfected females (i.e., genetic females) gained a mating advantage over infected females (feminized genetic males). In choice experiments, they discovered that males mate more often and for longer with uninfected than with infected females (Moreau et al. 2000). These differences resulted in higher fertilization rates for uninfected females. Observations of the copulation behavior showed that the differences were due to two factors. First, males interact more with uninfected than infected females. Second, feminized females do not have the full repertoire of behavioral responses to males during premating interactions. The result is that uninfected females have greater reproductive success than *Wolbachia*-feminized females and, consequently, the prevalence of the *Wolbachia* is depressed.

Host Coloration, Polymorphism, and Male-Killing

Biological Coincidences?

Before leaving the subject of the impact of male-killers on mating preferences, I want to go off on a speculative tangent. I do not believe much in biological coincidence. Therefore, when I see a biological pattern that is repeated several times, I tend to believe that there must be some evolutionary reason for the pattern. This is more the case when part of the pattern appears to present a biological conundrum. By that I mean that at least part of the pattern involves an observation that is at odds with current evolutionary theory. My bias is to believe the observation rather than the theory that it appears to contradict. Here then, I would like to focus on a group of related observations.

1. A high proportion of the species infected by male-killers is aposematic (e.g., ladybird beetles, milkweed bugs, acraeid butterflies); that is, they have true warning colors, being both brightly colored and unpalatable to many predators.
2. Most species with aposematic colors are monomorphic. This is rational because the whole idea of true warning coloration is to allow inexperienced predators to learn that a particular color pattern is associated with unpalatability. Evolutionary theory therefore predicts that members of aposematic species should all look the same.
3. A few species that have warning colors and have been shown to be aposematic are known to be highly polymorphic for color patterns, contradicting theoretical expectation (Plates 6 and 7).
4. Every one of the highly polymorphic aposematic species that has been assayed for male-killers has been found to harbor at least one.

As I say, I do not believe in biological coincidence, so my question is this: is there a relationship between the presence of male-killers and promotion of color pattern polymorphism in aposematic species?

Male-Killers Are More Likely in Species with True Warning Colors

Before addressing that question, it is necessary to point out that the prevalence of male-killers in aposematic hosts is certainly not coincidental. The most common reason for the spread of male-killers is through resource reallocation favoring the sisters of killed males. Benefit accrues either directly through the consumption of the dead males by female siblings, or through reduced competition between infected siblings. Such benefits are far more likely to result if siblings are in close proximity and aposematic species have a much greater tendency to aggregate than species that use other types of color pattern defense. This is because there is an element of kin selection in most models of the evolution of aposematism. Aggregating relatives with a particularly bright and memorable color pattern and unpalatability benefit from one another if an inexperienced predator tests any one of them. So male-killing is more likely to spread in aposematic species than in others because the advantage of male death is more likely to go to infected than uninfected members of the population. Said the other way around, the high proportion of species harboring male-killers that are aposematic suggests that, in most cases, resource reallocation is the driving force behind the successful invasion or evolution of male-killers. That said, because aggregating at high density must be considered to be selectively advantageous for aposematic species, any fitness compensation accruing to male-killer–infected females would have to outweigh the disadvantage of being in smaller family groups after male death. This would be in addition to the advantage needed to balance the loss of male progeny.

A second reason why male-killing may be more common in aposematic species than in others involves one of the risks of bearing an efficient male-killer. A host that harbors a male-killer with very high vertical transmission runs the risk that lack of males may seriously depress populations, possibly leading to extinction. However, if adults aggregate at a time when mating may take place, as is the case in many species of ladybird beetle, the impact that the biasing of the sex ratio toward females may have on fertility, through reduced mating opportunities for females, may be minimal.

There is a third possible link between male-killing and aposematism. A male-killing strategy may evolve in bacteria that play a direct mutualistic role in the defenses of their aposematic host, at least in some cases. Some distasteful organisms manufacture their own defensive chemicals while others collect them from the tissues of the plants they eat and store them. It is possible that some bacteria are able to aid their hosts in the collection and storage of defensive botanical chemicals. Inherited bacteria that aid their hosts in digestion are known. For example, termites and some cockroaches that eat wood can do so only because they are host to bacteria that can digest cellulose. It is possible that in some ecological circumstances, mutualistic gut bacteria may turn nasty, evolving male-killing as a more successful approach than simple mutualism. It is interesting that the closest known relative of the male-killers found in the ladybirds *Adonia variegata*, *Coleomegilla maculata* and *C. sinensis*, are Flavobacteria, as are the mutualistic gut bacteria of termites and some cockroaches.

Sibling Recognition by Color Pattern

We can now return to my other question: is there a relationship between the presence of male-killers and promotion of color pattern polymorphism in aposematic species? The pattern of association between color pattern polymorphism and male-killer presence is a strong one. Pairs of congeneric species, in which sex ratio distortion as a result of male-killers has been sought or would be expected to have been detected, were found for five genera (Table 7.1). In every one of these five genera, male-killers occurred in species with higher levels of color pattern polymorphism than those found in comparitor species that are not known to harbor male-killers. This suggests that the correlation is not merely a consequence of a study bias. The case of *Danaus chrysippus* is of particular interest in this context because not only can it be compared with *Danaus plexippus*, but also with itself, for over much of its range, it does not harbor a male-killer. Only in East Africa is it known to carry the male-killing *Spiroplasma* and only there is it polymorphic for its color pattern.

One reason for the association seems possible. It will benefit a male to mate preferentially with uninfected females. In highly polymorphic species, a male may be able to use his own color pattern to increase his chances of mating with an uninfected female. *Adalia bipunctata* may be used to illustrate this hypothesis.

Adalia bipunctata is highly polymorphic for the color patterns on its pronotum and elytra. The German entomologist Leopold Mader (1926–1937) described and named over a hundred different varieties, most of which have subsequently been recorded in southern England. All populations are polymorphic, with many of the forms occurring at low frequencies. Work on one of these rarer forms, the *annulata* form, has shown that it mates assortatively, preferring to mate with other *annulatas* rather than with the commoner typical form (O'Donald et al. 1984). This is not an instance where sibling proximity is likely to help males find and mate with siblings, because, in Britain, individuals that reach adulthood in one year tend to disperse some distance both to and from overwintering sites before mating the following spring. Now imagine a male from an uninfected lineage who also happens to have the genes responsible for one of the rarer forms. If he preferentially mates with individuals that have the same color pattern and so the same gene, there is an increased chance that his partner is a descendent of the same uninfected matriline. Because the color pattern variants of *A. bipunctata* are controlled by autosomal genes inherited from both parents, and not mitochondrial genes, the association between lack of male-killer infection may not be maintained for long if males with the color pattern variant sometimes mate with infected females. However, as most of the color patterns of two spots are controlled by a complex, multiple-allelic supergene, new variants will regularly be generated at low frequency within popula-

TABLE 7.1
Within genus comparison of male-killing presence and color pattern polymorphism in aposematic species. In each species mentioned, a minimum of 30 matrilines were assayed for the presence or absence of male-killing endosymbionts using phenotypic indicators (egg-hatch rates, larval survival rates, progenic sex ratios).

Species With Male-Killers	*Species Lacking Male-Killers*
Harmonia axyridis: male-killer present in all 16 populations sampled: highly polymorphic	*H. conformis*: no male-killer in 3 populations sampled: all samples monomorphic
Propylea japonica: male-killer present in single population sampled: highly polymorphic	*P. 14-punctata*: no male-killer in 3 populations sampled: polygenic variation; less variable than *P. japonica*
Coccinula sinensis: male-killer present in single populations sampled: melanic polymorphism	*C. 14-pustulata*: no male-killer present in two populations sampled: all samples monomorphic
Acraea encedon/encedana: male-killer present in >20 populations sampled: highly polymorphic	Other *Acraea* species: no confirmed male-killer present (but see *Acraea eponina* p. 117): most species monomorphic or showing less variation than *A. encedon/encedana*
Danaus chrysippus: male-killer present in more than 6 East African samples: highly polymorphic in East Africa	*D. chrysippus*: no male-killer outside East Africa: monomorphic outside East Africa. *D. plexippus*: no male-killer reported: monomorphic.

tions due to recombination within the supergene. This system is somewhat analogous to the use made by mice of variations in the major histocompatability complex that allows preferential mating with cousins. The subtle point in the *A. bipunctata* male-killer scenario is that if a new color pattern variant arises in an infected female, the male-killer she bears does not benefit from the assortative mating habit because she has no male relatives to mate with.

SEXUALLY TRANSMITTED DISEASE AND THE SEX RATIO

Adalia bipunctata may be used to illustrative another consequence of male-killing and the sex ratio bias that it produces. The sex ratio within a population has a strong impact on the way in which sexually transmitted diseases (STDs) may spread within a population. Assume that females control the number of matings that they engage in, at least until males become so scarce that females ready to mate cannot find a partner. In a population that is biased toward females, the average number of partners that each male has will increase, as will his chance of contracting an STD. As the prevalence of the STD increases in males, the prevalence will also increase in females. So in populations of promiscuous species, such as *A. bipunctata*, in which both sexes mate with multi-

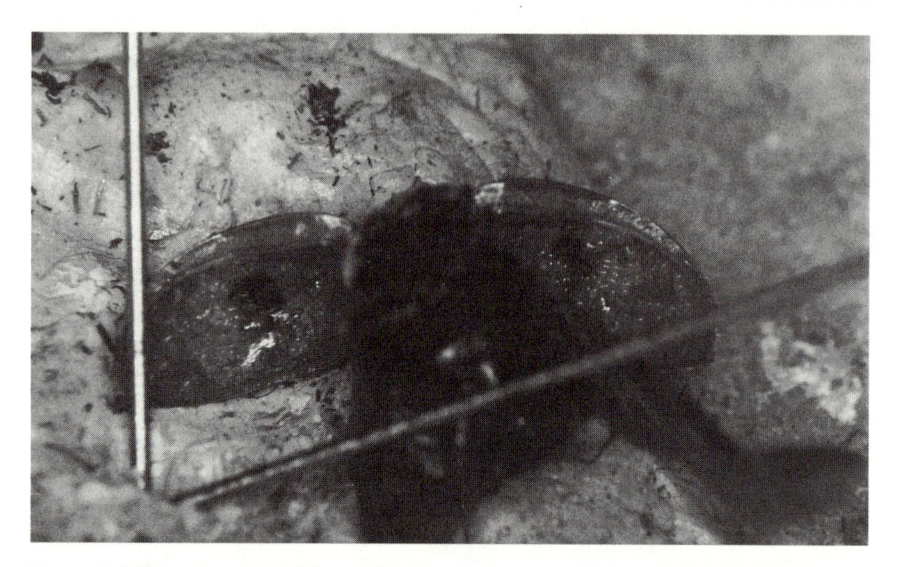

Figure 7.3 The mite *Coccipolipus hippodamiae* is a sexually transmitted disease of a number of species of ladybird. Females feed with their jaws embedded into the underside of the host's elytra. Larvae migrate onto other hosts when the hosts copulate.

ple partners, distorted sex ratios will increase the prevalence of sexually transmitted diseases, as long as the number of partners available to females is not too limited due to the scarcity of males. Few studies on the incidence of sexually transmitted diseases in invertebrates have been conducted. However, *A. bipunctata* bears both a collection of four male-killers, generally each at fairly low frequencies, and two STDs. These STDs are a mite called *Coccipolipus hippodamiae* (Figure 7.3) and a fungus of the genus *Laboulbeniales*. The interactions between the prevalence of the male-killers and the dynamics of the STDs certainly deserve attention.

Male-Killer Suppressors

Current sex ratio theory argues that, in the absence of local mate competition, biases in the sex ratio toward one sex or the other should impose selection for the production of the rarer sex (Fisher 1930; Hamilton 1967). We have already seen that in species that carry a male-killer, population sex ratios can become severely female biased. Selection in favor of male production should be intense in such populations. This selection will result both from the advantage of producing the rarer sex, which will have greater reproductive potential, and from efforts by the nuclear genome to resist the mortality imposed on it by the cytoplasmic genome in the form of the male-killers. As we discussed in chapter 4, autosomal genes that suppress female sex ratio–distorting phenotypes are known for both cytoplasmic male sterility in plants, sex chromosome meiotic drive in true flies, and the feminizing *f* factor in *Armadillium vulgare*. They are also suspected in the woodlouse *Porcellinoides pruinosus* that hosts a feminizing *Wolbachia*.

Populations that are female-biased due to invasion by male-killers may be subject to the evolution of either nuclear suppressors of male-killers, or mechanisms that bias the sex ratio toward males. The latter arise from distortion of either the primary or the secondary sex ratio toward males, as long as the loss of female offspring is offset by increased survival chances for male progeny. Theoretically, this is plausible only when biased male production is strongly beneficial due to highly female-biased population sex ratios. There is some evidence of male biases in some families of *A. encedon*, *A. encedana*, *C. sinensis*, and *H. axyridis* that would fit this criterion.

Were nuclear suppressors of male-killers to evolve, they could act in a variety of ways. They might kill the male-killer; reduce its vertical transmission efficiency by suppressing its replication rate; or prevent the symbiont from killing males, either by stopping the male-killer from detecting that it is in a male rather than a female host, or by interfering with and preventing the act of killing males. Recent work has revealed the first example of a male-killer suppressor.

Matsuka et al. (1975) first reported all-female families resulting from male-killing in the Asian ladybird, *Cheilomenes sexmaculatus*. The male-killer, which is transmitted in eggs, is horizontally transferable in hemolymph by micro-injection and curable by both high temperature and tetracycline treatment. Recently, 16S rDNA analysis of two female matrilines from Tokyo, has identified the male-killer as a γ-proteobacterium with closest similarity to a secondary endosymbiont of *Bemisia tabaci* (97% identity) and to a number of bacteria of the genus *Yersinia* that includes *Yersinia pestis*, the causative agent of the plague (Majerus 2001).

Crosses of progeny from these two SR lines, and males from a variety of N lines, suggested that the line from which males used as mates were drawn had an effect on the sex ratios produced. To test this possibility, males were swapped between females (Figure 7.4), and males from different lines were mated sequentially with SR females. The sex ratios of progeny produced after each mating were assessed, and some of the progeny were assayed for the presence of the male-killer. The results led to the hypothesis that some, but not all, males carried a factor, acting through sperm or some other element in the ejaculate, that inhibits the male-killing action of the bacterium. A single dominant nuclear gene could explain the initial data.

Experiments designed to determine the nature of the suppressor system confirmed that the gene is a single, dominant, autosomal gene. It is inherited in a typically Mendelian fashion, being transmitted through both females and males. The suppressor does not affect the inheritance of the male-killer in female hosts. Females from male-killer–rescued matrilines produce female-biased sex ratios when crossed with males not carrying the suppressor. Furthermore, molecular genetic analysis showed that rescued males carry the male-killer, but do not transmit it to their progeny. In this case, the suppressor gene acts as a rescue

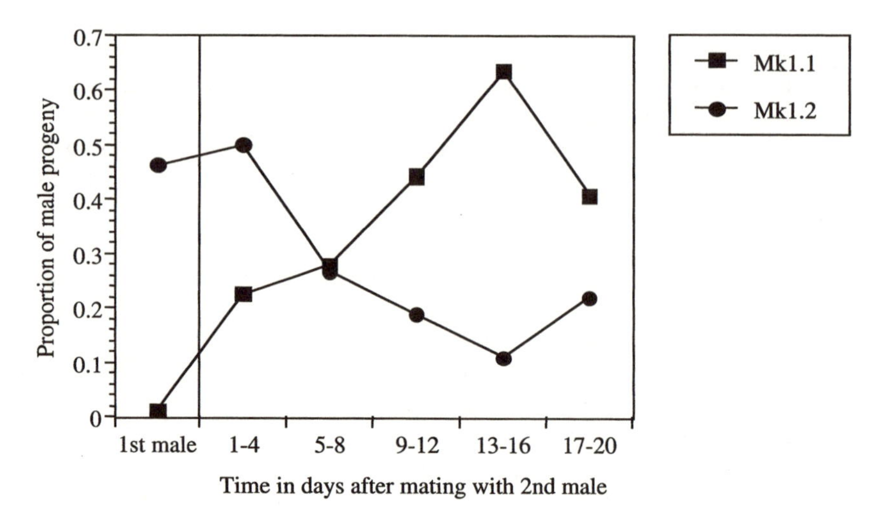

Figure 7.4 The effect of swapping males of *Cheilomenes sexmaculatus*. Females Mk1.1 and Mk1.2 were sisters and both carried a male-killing γ-proteobacterium. The first male to mate with female Mk1.1 was homozygous for a male-killer susceptible allele. The first male to mate with Mk1.2 was homozygous for the male rescue allele. Following male swapping, the progenic sex ratios produced by the two females changed significantly.

gene for males: it does not kill the bacterium or reduce its vertical transmission efficiency (Majerus 1999).

Male-Killer Rescue Genes and Intragenomic Conflict

The discovery of a nuclear gene that rescues males from the killing effects of a maternally inherited bacterium agrees with expectations from both sex ratio and intragenomic conflict theories (chapters 2 and 4). Selection favoring nuclear suppressor genes will be a direct consequence of female biases in the sex ratio. As mentioned previously, autosomal genes that act against sex ratio distorters have been recorded in isopods infected with feminizing *Wolbachia* and the *f* factor (p. 126). In *A. vulgare*, the main effect of such genes is to reduce bacterial transmission to progeny. By contrast, in *P. pruinosus* autosomal genes are conjectured to prevent the feminizing effect of *Wolbachia*. This is similar to the situation in *C. sexmaculatus*.

Most models that have investigated the evolutionary interactions between sex ratio distorting symbionts and suppressors are based on the assumption that the suppressor will kill the symbiont or reduce its vertical transmission. The dynamics of a male rescue gene may be altogether different and will depend on its cost, if any, in the absence of the male-killer. A male-killer in the presence of a male rescue gene should be selected against due to the cost on hosts of carrying the male-killer and the lack of fitness advantages to infected females resulting from male death. Such a male-killer, then, faces several alternative fates. It may be selected to extinction. It may become polymorphic, the prevalence of the male-killer then being affected by its transmission efficiency, the level of fitness compensation, and the costs to hosts of both it and the rescue gene. The male-killer may also evolve mechanisms to circumvent the rescue gene by evolving a different mechanism to kill males, as appears to be the case with the double feminizing effect of *Wolbachia* in *A. vulgare* (p. 125). Finally, it may reduce its cost on hosts, becoming costless or even beneficial. This final option attests to the assertion that cytoplasmic male-killers are an exquisite testing ground for theories of the evolution of virulence (Hurst et al. 1997).

Assessment of the distribution and frequencies of the male-killer and the rescue gene across the geographic range of *C. sexmaculatus* would certainly be valuable, as would work to assess whether the rescue gene imposes a cost on bearers, or the male-killer imposes a cost on infected females. Furthermore, molecular investigation of ladybirds with ecological and behavioral characteristics that should make them prone to male-killer invasion, but in which searches for male-killers using phenotypic assays have proved negative (e.g., *Hippodamia convergens*, *Harmonia conformis*), may reveal the presence of beneficial symbionts that constitute a peaceful resolution of an evolutionary arms race between a male-killer and a suppressor system.

Male-Killer Suppressors and the Presence of More Than One Male-Killer

The male rescue gene described above is the first male-killer resistance gene to be discovered. Its discovery is timely, for it may help explain one of the contradictions between theory and observation involving male-killing. Classical models of male-killers, based on the vertical transmission, costs to female hosts and levels of fitness compensation of male-killers, predict that long-term coexistence of multiple male-killers within a single host population is not possible. Yet this prediction is undermined by the finding of four different male-killers (a *Rickettsia*, a *Spiroplasma*, and two *Wolbachias*) in a single small sample of *A. bipunctata* from Moscow (Majerus et al. 2000). Noting this contradiction, James Randerson et al. (2000) have shown theoretically that two male-killing symbionts can coexist if the host evolved partial resistance to the stronger of the two male-killers (the one that would outcompete the other). The Randerson model does not provide an explanation for the multiple male-killers in Muscovite *A. bipunctata* because it makes a number of oversimplifications and assumptions that make it inappropriate to aphidophagous coccinellids. Furthermore, they assume that suppressor genes would have antibiotic rather than male-rescue characteristics. However, their result does give some hope that more sophisticated and appropriate models might permit coexistence of more than one male-killer, and so resolve the conundrum of multiple male-killers, in *A. bipunctata*.

It may also be noted that an alternative possibility is that the male-killers in *A. bipunctata* in Moscow are presently at a dynamic, non–frequency dependent equilibrium. Were this to be the case, the four male-killers would coexist at frequencies that fluctuate as a result of changes in environmental parameters that impact on both the male-killers' vertical transmission efficiencies and the resource advantages accruing to female progeny infected with the various male-killers. Differences in the relative fitnesses of the male-killers may change within a season or between seasons in an unpredictable way, as climatic and other factors that influence aphid demography vary. Here the polymorphism is not protected and a particular male-killer may be lost from a particular population following a protracted period when it, or its hosts, are consistently selected against compared to the other male-killers, or their hosts and uninfected hosts. However, such losses are likely to occur very rarely, since the decline in male-killer prevalence due to inefficient vertical transmission will be much slower than its increase when aphid scarcity causes the fitness compensation from male-killing to increase. Should rare loss occur, it may only be temporary, the lost male-killer later reinvading as a result of migration from other populations or by interspecific horizontal transmission from another host.

Male-Killers and Mitochondrial DNA

Different sex ratio distorters may impact on their carriers in different ways, depending on the mode of inheritance of the distorter and the genetic system of the host. We have already seen that driving sex chromosomes lead to the evolution of responder genes, while in some populations of *A. vulgare*, feminizing *Wolbachia* have effectively usurped the sex determination system. The antagonistic interactions between ultraselfish genetic elements and their hosts can lead to arms races of manipulative strategies and countermeasures.

Some sex ratio distorters will also have an impact on their host's genetic makeup, simply as a result of the symbiont's mode of inheritance. Taking male-killers as an example, Rufus Johnstone and Greg Hurst (1996) examined the effect of an invading male-killer on the mitochondrial DNA (mtDNA) of a host species. The critical point here is that bacterial male-killers and mitochondria are both cytoplasmically inherited. Initially, let us allow that a single host individual becomes infected with a male-killer. If this male-killer spreads, for whatever reason, it will drag the mitochondrial type (call it mitotype A) of its first host with it. Mitotype A will thus increase in frequency as the male-killer does. Furthermore, if the male-killer's vertical transmission efficiency is less than perfect, some of the progeny produced by females of this matriline will lose the bacterium. In this way, mitotype A will spread into the uninfected portion of the population. The proportion of the uninfected population that are descendants of the male-killer matriline that have lost the bacterium because of its imperfect vertical transmission will increase by a small amount each generation. Eventually, all hosts will carry mitotype A, whether they are infected with the male-killer or not, because all will be direct matrilineal descendants of the female that was first invaded by the male-killer. Variation in the mitochondrial DNA of such a population will thus be severely reduced. In practice, not all mitochondria of a host population are likely to be identical for two reasons. First, it is not necessarily going to be the case that the male-killer will invade just a single host female. Each separate invasion event will increase the mtDNA variability of the population, assuming that the mitotypes of the invaded hosts are different. Second, during the protracted period that it will take for a single mitotype to spread throughout a population, mutations will occur in the mitotype, gradually restoring at least some of the mtDNA variability.

This effect on mtDNA variability has important consequences for those who use DNA variability as a tool to study the magnitude of genetic changes between populations and the level of gene flow between populations. Thus, for example, much work on population structure, gene flow, and taxonomy is now conducted through examining variation at the level of DNA sequences. In this work, mtDNA sequences are frequently used. The presence now, or in the past, of a male-killing bacterium in a lineage under investigation will have a profound effect on the data obtained and its interpretation. This stresses the need for the use of both mitochondrial and nuclear DNA analysis in studies of genetic bottlenecks and gene flow.

A second possibility is that a male-killer will have effects on the nucleotide sequences of the mtDNA (Majerus and Hurst 1997). As male-killers are detrimental to their hosts, the hosts will be exposed to selection pressures to combat the male-killer. One mechanism of achieving this would be by reducing the vertical transmission efficiency of the male-killer. To achieve such a reduction, a host might increase the replication rate of its germ line cells, to reduce the probability of each daughter cell receiving at least one copy of the bacterium. Here, then, the idea is that the germ line cells of the host multiply at a faster rate than can be achieved by the bacterium. When cells divide, either mitotically or meiotically, mitochondria also have to be replicated. Most cells of higher organisms that lack mitochondria simply die. Faster cell division may produce selection for a reduction in the size of the mitochondrial genome. Selection may favor loss of those parts of the mitochondrial genome that do not affect the function of gene products. There may also be effects on the sequence of DNA bases, apart from such deletions, to "accommodate" such changes. If this occurs, it will have implications for the use of branch lengths as estimators of genetic distance. Faster evolution of mtDNA may occur in male-killer–infected lineages.

The Evolution of Virulence

There has been a long-held general view that it does not pay for disease-causing parasites to be too virulent, and, indeed, there is an expectation that, over time, selection should favor lineages that are less virulent and do less damage to the host on which they are reliant. The basis of this argument was group selectionist: if they were too virulent, parasites might bring about the extinction of their hosts and so themselves. More recently, it has been accepted that, in many circumstances, more virulent mutations will outcompete less virulent strains and so spread, so that there is no general evolutionary trend from virulence to avirulence.

In the more specific instances of microorganisms that live in the cells of their hosts and are exclusively vertically transmitted, there is a similar argument: that because of the complete reliance on their host, these should evolve to be beneficial to their hosts and so themselves. This argument is correct for many symbionts, but ignores those microbes that manipulate host reproduction to their own ends.

In this context, the case of male-killers is peculiar, for selection should favor increased virulence in male hosts and decreased virulence or even beneficial effects in female hosts. This is easy to understand in the ladybirds, where spread of the male-killer is dependent on the consumption of male-killed eggs by male-killer–bearing females. The advantage that accrues to these females is greatest when male embryos are killed at a very young age, and so have used up very little of the soma of their egg. More virulent strains of male-killers will gain if that virulence is manifest in earlier killing, through the kin selection resulting from resource reallocation. Conversely, as these symbionts are transmitted through females, any harm they do to female hosts, is, in effect, harming themselves. Selection should thus favor strains that have no effect, or a beneficial effect, on female hosts. By definition, male-killers have different virulence levels in the sexes of their host, killing males but not females. Furthermore, it has not been demonstrated that the virulence of any of these microorganisms has changed since invading a host. Because the act of male-killing seems to be dependent on the density of the male-killer in some instances, rapid replication of the microbe, which is usually associated with increased virulence, may be more strongly selected for in male hosts. It would thus be valuable to assess replication rates of bacteria in male and female eggs. If differences were found, with replication more rapid in males, this would suggest that selection had indeed caused a sex-specific alteration in virulence.

The situation is perhaps not so clearcut as this, since selection might also favor increased replication in females if this increased the vertical transmission of the symbiont. Yet increased density of the bacterium within a female host may have negative effects on her, such as lower fecundity or longevity, which would impact adversely on the symbiont. It may be, therefore, that within female hosts, there is a tradeoff in terms of the replication rate of the bacterium: if too high, the cost to the host, and so themselves, is too great; if too slow, the vertical transmission efficiency of the symbiont drops.

Clade Selection

Still on the subject of virulence, it is interesting to note that while *Wolbachia* that cause CI typically have a vertical transmission efficiency close to 100%,

Wolbachia that cause feminization and male-killing do not always have such a high vertical transmission efficiency. Similarly, the vertical transmission efficiency among other male-killers is quite variable, with values below 80% having been recorded. It is seductive to think that these rather low transmission rates may be the result of a virulence/transmission tradeoff. There is, however, another possibility.

In one sense, the best (i.e., "strongest" *sensu* Randerson et al. 2000) male killer is the one with the highest basic rate of increase. That is to say it has perfect vertical transmission, kills all male hosts, and imposes no cost on female siblings. Because the best male-killer will displace others, one might expect that male-killers with near-perfect vertical transmissions would predominate. Some male-killers, notably the *Spiroplasmas* from *A. bipunctata* and *H. axyridis*, the Flavobacterium from *C. sinensis*, and one of the *Wolbachias* from *A. bipunctata* have vertical transmission rates of over 99%. However, most have lower rates, ranging from about 80% to 95%.

These lower transmission rates could be a result of selection acting at an altogether higher level than that normally considered—the level of the population or even species. In populations of hosts harboring male-killers with near-perfect vertical transmission, efficient male-killing ability, and low cost to female hosts, a male-killer could spread to such high prevalence that males become extinct, with extinction of the host population following very shortly afterwards. In the case of *Wolbachia* in the *Acraea* butterflies, this type of selection might be a reality. If it is, then here those male-killers that persist longest may be those that have lower vertical transmission, or lower male-killing abilities, for in these at least some males will be produced each generation.

Here, then, male-killers with imperfect vertical transmission may have greater evolutionary longevity than those with near-perfect vertical transmission. They will persist for longer simply because the likelihood that they will drive their host to extinction (through lack of males) is reduced. Selection is then acting at the level of the whole population or group, rather than at the level of the individual.

Parthenogenesis: An Escape Route

There are two final ways that a host infected by a strong male-killer may avoid the loss of half her progeny, namely her sons. Both involve her not producing any sons, but the two differ in one important respect: in the first, males are still required to fertilize females; in the second, males become completely redundant.

To understand the first pathway, imagine a host population infected at high prevalence with a strong male-killer. To help the imagination, let us assume that the male-killer is in a butterfly that lays its eggs singly rather than in clutches, as is the case in *D. chrysippus*. If infected females produce equal numbers of female and male eggs, half their reproductive expenditure is wasted. But if an infected female could lay just female eggs, she would avoid this wastage. The ability to lay just female eggs would depend on her being able to control the sex ratio of her offspring in response to the presence of the male-killer. As females would be the commoner sex in such a population due to the high prevalence of the male-killer, uninfected females would be selected to produce predominantly the rare sex and so have mainly or exclusively sons. Currently there is no proof that this pathway has been followed in any known instance. However, data from two species of butterfly suggest that uninfected females may produce an excess of males. In *A. encedana*, data for twelve families in which the females were not infected by male-killing *Wolbachia* give a sex ratio of progeny of 163 males:118 females (Jiggins et al. 2000b). The chance of this ratio occurring by chance is less than one in a hundred. Significantly male-biased families have also been reported from the ladybird *C. sinensis* (Majerus and Majerus 2000a).

The second pathway allows a complete escape from the wastage of producing males that are killed by a symbiont, for it is to evolve parthenogenetic reproduction. Whether parthenogenesis has ever evolved as an escape route for a species that is devoid of males because of the presence of a male-killer is an open question. A recent study of male-killing in a bupresid beetle, *Brachys tessellatus*, suggests that there may be a link.

Lawson et al. (2001) have shown that the sex ratios found in this leaf-mining beetle are highly female biased, with sex ratios ranging from 1.3 females to 6.0 female per male. Many sexually reproducing females produce all-female or largely female families and high levels of egg mortality. Antibiotic treatment led to an increase in both egg-hatch rates and the number of males produced. Molecular techniques showed that a *Rickettsia*, related to those identified causing male-killing in three species of coccinellid was highly associated with the production of female-biased families. Of particular interest to the issue of whether host species in which males became rare due to the spread of male-killers could evolve parthenogenesis as a "way out," is the finding that about 30% of females denied mating opportunities produced at least some viable eggs. Lawson et al. (2001) estimate that as many as 12.2% of eggs in the wild may be the result of parthenogenetic reproduction. They speculate that if males become sufficiently scarce that females regularly go uninseminated, parthenogenetic reproduction will be strongly favored. This then may be a case in which genetic conflict between a cytoplasmic genetic element, in the form of the male-killer, and nuclear genes is leading to selection for a major change in reproductive strategy (Juchault et al. 1993; Werren and Beukeboom 1998).

In *B. tessellatus*, it is not yet known whether parthenogenesis has resulted from selection due to male rarity or whether the *Rickettsia* causes parthenogenesis directly. If the latter were found to be the case, this would represent the first case of parthenogenesis being induced by a bacterium other than *Wolbachia*. *Wolbachia*-induced parthenogenesis is the subject of chapter 8.

CHAPTER 8

Parthenogenesis Inducers

> Come, you spirits
> That tend on mortal thoughts! unsex me here. . . .
> —William Shakespeare, *The Tragedy of Macbeth*

SUMMARY

Secondary asexual reproduction has evolved in a wide array of taxa. Only in 1990 did it become clear that parthenogenesis could be induced by intracellular symbionts. In all confirmed cases to date, symbiont-induced parthenogenesis is caused by *Wolbachia*. The aim of this chapter is to outline the research that led to this discovery and to describe some of the subsequent investigations of the phenomenon.

FORSAKING NEEDLESS MALES

The final way out for a species driven to the point of extinction by an efficient male-killer is to evolve an asexual lifestyle. Secondary asexual reproduction, or parthenogenesis, is widespread in higher animals and has evolved independently on numerous occasions, with a variety of different mechanisms being employed (p. 25). Most major taxa have a mixture of sexual and parthenogenetic species, with the sexual species generally, but not always, being in the majority (p. 35). In a few taxa, one or another mode of reproduction has a monopoly. Thus, in bdelloid rotifers, males have never been recorded, while all species of mammals reproduce sexually. The reason that mammals are never parthenogenetic (setting aside the birth of Christ) is a consequence of a secondary adaptation of sexual inheritance, called gene imprinting, that has only evolved in higher animals, such as birds and mammals. For some genes, it matters whether they have been inherited paternally or maternally. Some are only expressed if they are inherited from their father, others only if they derive from the mother. Copies of these genes from the alternative parent are switched off. Because these imprinted genes are essential to development, both a mother and father are essential, making a change to a parthenogenetic lifestyle impossible.

However, mammals are exceptional in lacking parthenogenetic species. Females of many species have forsaken males and, in some of these species, inherited microorganisms are involved.

Parthenogenesis-Inducing Microbes

Much of the biotic world indulges in sexual reproduction. However, a significant proportion of organisms reproduces asexually, either ancestrally, or secondarily when asexuality has evolved from sexuality. Secondary asexual reproduction involves the development of new individuals from unfertilized eggs. In diploid species, the evolution of parthenogenesis requires two steps. First, the egg has to be able to develop without the catalyst of sperm penetration of the egg. Second, the diploid number of chromosomes has to be restored in the haploid egg. In some organisms, however, only females are diploid, males being haploid and arising without fertilization. In these haplo-diploid species, development of male embryos is initiated without the stimulus of fertilization, and other mechanisms of initiation of embryonic development have evolved. These species thus only need to take one evolutionary step, the restoration of the diploid number in females, to become fully parthenogenetic. It is for this reason that full parthenogenesis, the development of all progeny of a female without fertilization, is much commoner in haplo-diploid species than in fully diploid species.

A variety of types of parthenogenesis exist. Normal haplo-diploid species, in which females result from fertilized eggs and males from unfertilized eggs, are called *arrhenotokous*. In *deuterotoky*, both males and females arise from unfertilized eggs, while *thelytoky* involves production of just female offspring. The production of only female offspring, without the need for males, raises the possible involvement of cytoplasmic symbionts whose interests are favored by female hosts that can vertically transmit them. This involvement has been realized in a number of haplo-diploid groups (Table 8.1), with those in the Hymenoptera having received most attention.

The Wasp and the Wolbachia

Many small wasps are parasitoids of other insects. Some of these species have been used as biological control agents of pest insects. This is particularly true of parthenogenetic species of parasitoids, because their rate of reproduction and so their efficiency in pest control is greater than that of similar sexual species. During the first half of the 20th century, there were a number of reports of the production of males in normally thelytokous species that were being monitored in biological control programs (e.g., Perkins 1905; Flanders 1945). From these reports, the production of males appeared to occur during high summer or when wasps were reared under high temperatures. Later, systematic work to examine the effect of temperature in a variety of thelytokous Hymenoptera showed that in many but not all species, submitting females to temperatures in the high twenties or thirties (Celsius) led to the production of male progeny. Thus, for example, in *Ooencyrtus submetallicus*, if females were exposed to a temperature greater than 29.5°C during development and adult life, they only produced

TABLE 8.1
Arthropods in which *Wolbachia*-induced parthenogenesis has been reported.

Order	*Species*	*References*
Hymenoptera	Over 30 species reported	See Stouthamer 1997; Stouthamer et al. 1999 and references therein.
Collembola	*Folsomia candida*	Vanderkerckhove et al. 2000
Thysanoptera	*Franklinothrips vespiformis*	Arakaki et al. 2000
Acaria	*Bryobia* sp.	Weeks and Breeuwer 2000

sons (Wilson and Woodcock 1960a, 1960b). Similar results were obtained for a variety of other species from a range of genera, including *Trichogramma* (Bowen and Stern 1966).

The first stringent evidence that thelytoky in parasitoid Hymenoptera sometimes involved cytoplasmic rather than nuclear factors was obtained by Stouthamer et al. (1990a). Using *Trichogramma pretiosum*, attempts to place the nuclear genome of a thelytokously parthenogenetic line, via males produced by temperature treatment, into a conspecific arrhenotokous line over nine generations, rather surprisingly failed. Subsequent treatment of a thelytokous line with antibiotics (tetracycline hydrochloride, sulphamethoxazole, rifampicin) or with high temperatures (about 30°C), produced sons among the progeny of this previously all-female line (Stouthamer et al. 1990b). This led to the deduction that some sort of cytoplasmic microorganism, probably a bacterium, caused thelytoky in this species. Microscopic examination using lacmoid staining showed that a bacterium was present in eggs of some thelytokous lines, but not in antibiotic- or temperature-treated lines, or in arrhenotokous lines (Stouthamer and Werren 1993). DNA sequence analysis of the 16S rRNA gene revealed the presence of a *Wolbachia* (Rousset et al. 1992). *Wolbachia* have subsequently been found in a wide range of thelytokous Hymenoptera (Stouthamer 1997).

The Diversity of Symbiont-Induced Parthenogenesis

Evidence of microbe-induced parthenogenesis (PI) has now been revealed in over 30 species of parasitoid wasp and in a small variety of other Arthropod groups (Table 8.1). In every case except one, host species are haplo-diploid. Furthermore, in those instances in which the microbe involved has been identified, it is, without exception, *Wolbachia*. Microbe-induced parthenogenesis thus appears to parallel CI in being exclusively caused by *Wolbachia* and differs from feminization and male-killing where a variety of microorganisms are involved.

Outside the Hymenoptera, *Wolbachia* PI has recently been reported from three groups. Thrips are small, elongate insects of the order Thysanoptera. Reproduction in this group is usually by arrhenotoky, but thelytoky also occurs in some species. The predatory thrip, which preys on mites, leafhoppers, and whitefly, was investigated by Arakaki et al. (2000). They found all-female populations in Japan and showed that the females were host to two strains of *Wolbachia*. In captivity, females produced only female progeny for ten generations. However, when given tetracycline or submitted to heat treatment, males were produced. These males were found to be fully functional, producing mobile sperm and mating with females. However, *Wolbachia*-cured females did not utilize the sperm transferred by these males during mating for fertilization. Rather, all progeny produced by these cured females were male. Arakaki et al. conclude that *Wolbachia* causes thelytokous parthenogenesis in this thrip.

A similar case has been reported in a group of species of mites of the genus *Bryonia*. Thelytokous parthenogenesis predominates in these mites. Assaying six species for the presence of *Wolbachia*, Andrew Weeks and Hans Breeuwer (2000) found all to be infected. Females of three of the species were then treated with antibiotics and subsequently produced only male offspring, showing that the *Wolbachia* induces parthenogenesis.

A somewhat different pattern is seen in the springtail, *Folsomia candida*. Tom Vandekerckhove and his colleagues (1999) studied two thelytokous populations of this collembolan infected by *Wolbachia*. Asexual reproduction in this primitive insect is rather different from that in haplo-diploid species, for springtails are diploids. Females produce diploid eggs by gametic duplication, haploid eggs being aborted. Antibiotic treatment of females led to a lowering of the hatch rate of eggs from both populations. Examination of the density of *Wolbachia* in females from the two populations and in cured and uncured lines led Vandekerckhove et al. (2000) to conclude that there is a threshold density of *Wolbachia* that induces gametic duplication. Unmated females that lack *Wolbachia* above this threshold density produce haploid eggs that are aborted. The response to antibiotic treatment, a lowering of the hatch rate, is thus different in this diploid species from that seen in haplo-diploids where antibiotic treatment leads to all-male progeny.

Apart from the confirmed cases of PI *Wolbachia*, a report of a *Wolbachia*-like parasite in a stored product pest may extend the range. *Liposcelis bostrychophila* is a small insect of the order Psocoptera, which includes creatures such as book lice (Yusuf et al. 1999). A microbial parasite of this psocid was characterized by electron microscopy rather than DNA analysis, but has all the features of a *Wolbachia*. Although it has not been formally demonstrated that the bacterium causes the parthenogenetic reproduction typical of this species, it does seem likely, because closely related species of psocid that indulge in sex do not have the bacterium.

Horizontal Transfer of Parthenogenesis-Inducing Wolbachia

The *Wolbachia* that induce parthenogenesis do not form a single taxonomic group within the *Wolbachia*, but are intermixed with *Wolbachia* that affect hosts in other ways (e.g., CI and male-killing). This suggests that either PI has evolved independently several times in the *Wolbachia*, or that the genes that cause PI can be horizontally transferred between *Wolbachias*, perhaps through

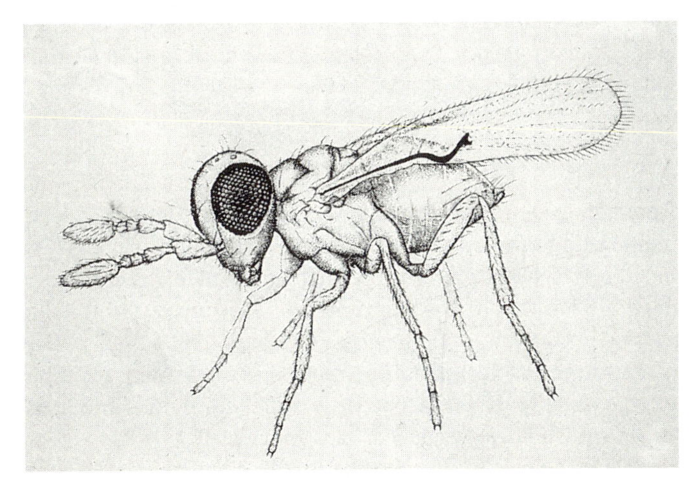

Figure 8.1 A trichogrammid wasp. (Courtesy of Richard Stouthamer.)

DNA exchange involving virus-like particles or plasmids. A third possibility is that the same *Wolbachia* causes different effects in different hosts; for example, PI in haplo-diploids, and CI or male-killing in diploids.

The PI *Wolbachia* in trichogramminid wasps, which are egg parasitoids of a number of insects, particularly Lepidoptera, are an exception to the rule that *Wolbachia* in closely related hosts are frequently very different. All *Wolbachia* reported from *Trichogramma* species come from a single taxonomic group, with no other known *Wolbachia* belonging to this group (Stouthamer et al. 1993; Werren et al. 1995b). However, phylogenetic analysis of a number of PI *Wolbachias*, from various species of these wasps, has provided evidence for frequent interspecific horizontal transmission of the symbionts. Within a *Trichogramma* species, different *Wolbachias* exist with no correlation between the phylogeny of the *Trichogrammas* and their various *Wolbachias* (Schilthuizen and Stouthamer 1997).

To investigate the regularity of horizontal transmission and its mechanism, Schilthuizen et al. (1998) studied two species of wasp, *Trichogramma kaykai* and *Trichogramma deion*, which coexist in the Mojave Desert in California (Figure 8.1, Plates 8a and 8b), and both of which carry PI *Wolbachia*. Arguing that horizontal transmission might be expected when female wasps of different species lay eggs into the same host, they analyzed the strains of *Wolbachia* in each species. The expectation was that *Wolbachia* strains would not be host-specific if interspecific horizontal transfer were common. However, from 40 PI strains of *T. kaykai* and 5 of *T. deion*, they found no evidence of horizontal transfer between the species. They thus concluded that horizontal transmission must be rare on an ecological time-scale. However, subsequent work investigating horizontal transmission from infected into uninfected lines of the same species has produced rather different results. Examining wasps from host eggs that had been infected by both *Wolbachia*-bearing and *Wolbachia*-free wasps of the same species, Huigens et al. (2000) found evidence of horizontal transmission in up to 30% of such hosts.

How Is Asexuality Induced?

Studies of parthenogenetic organisms have shown considerable diversity in the way in which diploid offspring can result from unfertilized eggs. This diversity can be split into two basic groups. The first group involves meiotic modification, in which the number of chromosomes is not reduced during meiosis. The second involves postmeiotic modification, in which haploid cells with a single pronucleus are formed as normal through meiosis, after which the diploid number is restored, typically by the fusion of two haploid mitotic products.

In the cases of *Wolbachia*-induced parthenogenesis that have been studied cytologically, the latter of these two routes is utilized. In both infected and uninfected females, meiosis is the same, with a single pronucleus being produced. In uninfected females (arrhenotokous), arrival of sperm leads to fertilization and normal female development. If no sperm arrives, the haploid pronucleus divides mitotically, with the two sets of chromosomes formed migrating to opposite poles at anaphase so that two haploid nuclei result. Further cell division proceeds along similar lines, so that a haploid male is produced. In infected females, the process is the same up to anaphase of the first mitotic division. At this point, both sets of chromosomes migrate to the same pole, with the result that a single diploid mitotic product is produced. This chromosomal behavior is known as *gametic duplication*. The methods by which *Wolbachia* affect the biochemical and/or mechanical behavior of host chromosomes during anaphase of the first mitotic division are currently unresolved. An analysis of the behavior of chromosomes in springtails that harbor PI *Wolbachia* would certainly be valuable as a comparitor to haplo-diploid Hymenoptera because of the normal diploidy of springtails.

In Hymenoptera showing *Wolbachia* PI, the temperature to which larvae and pupae are exposed strongly influences the sex of progeny subsequently produced. For most species, at temperatures below 26°C, all progeny are female, while over 30°C only males result. At temperatures between these limits, males, females, and gynandromorphs may all be produced (Cabello and Vargas 1985). In some, the change from all-female to all-male offspring may be extremely abrupt. Thus, in *O. submetallicus*, a change in temperature of just 0.2°C produced a change from all-female to all-male offspring (Wilson and Woodcock 1960b).

In species in which gynanders are produced at critical temperatures, Stouthamer (1997) suggests that differences in the proportion of male and female tissues may be the result of *Wolbachia*-induced gametic duplication being delayed, that is, not occurring at the first mitotic division. This hypothesis accords with the observation of a positive correlation between increasing temperature and the proportion of male tissue. However, if true, one would expect that the minimum proportion of male tissue in gynanders would be two-thirds, and this is not the case, some individuals being predominantly female with only small anterior clusters of male cells. In addition, there appears to be a front/back polarity in the tissues that are male and those that are female. In general, anterior tissues are more likely to be male than posterior tissues. The reason for this polarity is unknown. Clearly, further work is needed to understand the mechanism of cellular sex determination and the influence of *Wolbachia* in these parthenogenetic species.

The Population Dynamics of Parthenogenesis Inducers and Their Hosts

In most species of Hymenoptera showing *Wolbachia* PI, thelytoky has been fixed. All members of the population harbor *Wolbachia* and all produce just female offspring unless subjected to abnormally high temperatures. However, in many of the *Trichogramma* species, and possibly a few other hymenopterans, both thelytokous and arrhenotokous individuals coexist, with levels of *Wolbachia* infection generally being low (reviewed in Stouthamer 1997).

The reasons for this distribution of polymorphism and monomorphism are not clear as yet. Polymorphism may result from a variety of factors. If vertical transmission is less than 100% efficient, some uninfected individuals will result. Nuclear genes that kill *Wolbachia* or reduce its vertical transmission may also lead to uninfecteds. Once uninfecteds occur in the population, their frequency may increase if there are negative fitness effects of *Wolbachia* infection on the host. These factors are not mutually exclusive, so that they may work in combination to produce the low *Wolbachia* prevalences reported from some species.

Unfortunately, empirical evidence relating to these factors is rather limited. Vertical transmission rates have only been assessed in laboratory cultures, where they tend to be high: 90% for *Trichogramma* species tested, although transmission rates appear to decline with host age. Strong negative fitness effects of *Wolbachia* infection have been demonstrated in some cases. These result from heavily reduced offspring production, revealed by comparison of genetically identical infected and uninfected lines of *Trichogramma* exposed to unlimited hosts (Stouthamer and Luck 1993). However, if such strong negative effects occur in field populations, it is difficult to see why the infection is not lost. It is probable that, in the field, host availability is limited, and under such conditions infected females produce more daughters than do uninfected females (Stouthamer 1993).

Finally, although the conflict between the cytoplasmic genome and the nuclear genome should impose a selection pressure that would favor the evolution of nuclear suppressor genes, evidence for the existence of such suppressors is sparse and circumstantial. The evidence comes from two instances, in different species, of segregation of virgin progeny into those that produced only females and those that produced only males (Rössler and DeBach 1972; Stouthamer 1997). Although more stringent evidence of the existence of nuclear suppressors of PI *Wolbachia* awaits detection, theoretical simulations of the equilibrium

TABLE 8.2
The different sex phenotypes of Hymenopterans with *Wolbachia*-induced parthenogenesis. The sex of an individual depends on the infection status of the mother, whether it was fertilized, and whether the *Wolbachia* was vertically transmitted to it.

Infection Status of Mother	*Infection Status of Progeny*	*Fertilization of Egg*	*Sex of Progeny*
Infected	Infected	Fertilized	Female
Infected	Uninfected	Fertilized	Female
Infected	Infected	Unfertilized	Female
Infected	Uninfected	Unfertilized	Male
Uninfected	Uninfected	Fertilized	Female
Uninfected	Uninfected	Unfertilized	Male

produced by such suppressors are encouraging. Models that assume that a suppressor imposes some cost on its carrier, when homozygous, produce rather robust equilibrium values for the prevalence of PI *Wolbachia* at around 10–20%. In this model, an equilibrium is maintained because the suppressor in haploid males can be transmitted through mating with infected females and this transmission offsets the cost of the suppressor in females.

It is interesting to note that the prevalence levels of PI *Wolbachia* in different populations and different species are bimodal. Either prevalence levels are relatively low (typically 10–20%) or *Wolbachia* are fixed in the population. Host populations that are polymorphic for *Wolbachia* infection at high prevalence have not been reported. This pattern may be an inherent property of the population dynamics of PI *Wolbachia*. Once prevalence of infection exceeds a critical level, other factors lead to further increases in prevalence, such that a positive feedback loop drives the infection to fixation. As *Wolbachia* prevalence increases in a host population, males become rarer because eggs of unfertilized infected females will be female rather than male. Males are only produced from unfertilized eggs of uninfected females and the unfertilized eggs of infected females that have not inherited the *Wolbachia* from their mother (Table 8.2). The only uninfected daughters that infected females produce are from eggs that fail to inherit the *Wolbachia* and have been fertilized. This number will diminish as the prevalence of the *Wolbachia* increases because males, and so fertilization, become progressively rarer. Once *Wolbachia* prevalence reaches very high levels, the only uninfected individuals produced will be males resulting from infected females due to the imperfect vertical transmission of the symbiont.

Effects of Parthenogenesis-Inducing Wolbachia *on Host Mating*

In *Wolbachia* PI species of wasp, males can be produced by antibiotic or high temperature treatments. Observations of the behavior of such males when faced with females have shown that mating rarely takes place. In most studies, males appear to show some courtship behavior and sometimes attempt to copulate. However, females fail to show appropriate responses to the males' advances. For example, in *Apoanagyrus diversicornis*, in which both arrhenotokous and thelytokous lineages occur naturally, arrhenotokous females will mate with either naturally occurring males (i.e., from arrhenotokous mothers) or experimentally produced males from thelytokous mothers. However, females from thelytokous lineages will not mate with males of either type (Pijls *et al.* 1996). This suggests that thelytokous females lose critical elements of their courtship and copulatory behavior more rapidly than do males.

There are two possible reasons for the loss of sexual behaviors. First, when a PI *Wolbachia* infection reaches high prevalence, due to the lack of males, there will be little sexual reproduction in the population. There will thus be no selection to conserve precisely genes for sexual traits. Consequently, mutations may accumulate in those genes that control sexual reproduction. As inappropriate mating is more critical for females than for males, because of the high investment females put into eggs compared to the investment of males in sperm, females are likely to have more genes for sexual traits than males. This may account for the more rapid loss of sexual behavior in females than males in thelytokous lineages (Stouthamer 1997).

Alternatively, or possibly in addition, some sexual traits may have energetic or temporal costs. If this is the case, then in populations in which sexual reproduction is rare or absent because of the lack of males, selection may actively act against the maintenance of sexual behavior (Pijls et al. 1996).

How Are Hosts Invaded?

It is clear that research into the evolutionary effects of PI symbiotic microorganisms is still in its early stages. Since the involvement of cytoplasmically inherited *Wolbachia* in PI was first discovered, achievements have included: the identification of over 30 *Wolbachia*-Hymenoptera systems; the finding of PI *Wolbachia* in a small but expanding range of other arthropod taxa; considerable phylogenetic analysis that has suggested interspecific horizontal transfer of symbionts; demonstration of intraspecific horizontal transfer from infected to uninfected lineages; and understanding of the cytological basis of such parthenogenesis. Theoretical considerations of the population and evolutionary dynamics of PI have also advanced considerably (see Stouthamer 1997). However, there are still many gaps in our knowledge. By what mechanism does *Wolbachia* affect the behavior of host chromosomes? How precisely do gynanders arise? Does *Wolbachia* impose negative fitness effects on its hosts in field situations? How widespread and frequent are horizontal transmission of *Wolbachia* (either intra- or interspecific) in the wild? Do suppressor genes exist and, if so, how do they operate? These questions provide a fertile field for future investigation.

What can be said from the cases reported to date is that PI by *Wolbachia* appears to be a manipulation that depends in large part on the genetic basis of reproduction of the host. In all cases except one, PI *Wolbachia* hosts are haplodiploid. In the one exceptional case, that of the springtail, *Folsomia candida*, diploidy is restored in unfertilized eggs by gametic duplication. It would certainly be valuable to assay other parthenogenetic diploid species that produce diploid zygotes by gametic duplication for *Wolbachia* or other clades of inherited endosymbionts.

Is Parthenogenesis an Escape Route?

I introduced this chapter by suggesting that in populations of a species that became severely female biased due to the presence of male-killing or feminizing inherited microorganisms, parthenogenesis might be an escape route (p. 189). However, with the exception of the intriguing case of the bupresid beetle *Brachys tessellatus* (p. 188), this proposition does not appear to fit the evidence currently available, at least for most species. Evidence that sex ratio distortion caused by feminizers or male-killers subsequently led to the evolution of PI would require the finding of symbiont-induced parthenogenesis in host species susceptible to these sex ratio manipulations. Alternatively, the taxonomic diversity of PI and feminizers or male-killers might be similar. The diversity of PI and the diversity of their hosts both lead to the conclusion that PI is not a response to feminization or male-killing for most organisms. From the point of view of the symbionts, all known cases of PI are due to *Wolbachia*, while feminization and male-killing are both caused by a taxonomically diverse range of microbes. With respect to hosts, PI has mainly been recorded in haplodiploid species, with only one case known from a diploid insect. Yet feminization and male-killing are largely phenomena of diploid species.

The evolution of parthenogenetic reproduction from sexual reproduction in a diploid requires both the initiation of zygote development without the catalyst of fertilization and restoration of the diploid number. These are both difficult evolutionary steps. The ability to achieve both at the same time appears to be a trick that is one step too far, even for the wonderfully manipulative skills of *Wolbachia*.

CHAPTER 9

Sex Ratio Distorters

IMPLICATIONS AND APPLICATIONS

It has long been an axiom of mine that the little things are infinitely the most important.

—Sir Arthur Conan Doyle, *A Case of Identity*

Summary

In this final chapter, I return to the question of the evolution of sex, revisit Darwin's entangled bank, and consider the importance of sex ratio distorters. In particular, the implications for evolutionary genetics of the interactions that *Wolbachia* and other microorganisms that pursue similar ultraselfish strategies have with their hosts are discussed. The question of whether *Wolbachia* is unique among intracellular symbionts is then addressed, before the applications that *Wolbachia* and sex ratio distorters may have in pest control are considered. Finally, a number of possible future discoveries are suggested.

Musical Chairs

My wife, Tamsin, and I have three children, a daughter and two sons, all between the ages of 8 and 11. Their existence in my life is possibly the reason why a recent news item on British radio caught my attention. The item concerned a report that advised parents to stop young children from playing party games such as musical chairs and hide-and-seek. The thinking behind this advice was that such competitive games encourage aggression and conflict. I fear that the well-intentioned authors of the report will find, even were their advice to be heeded, that the reduction in aggression resulting would be so minimal as to be unmeasurable. The aggression of our children is built in. It should be unacceptable to blindly blame human antisocial behavior on the genetic makeup of the individual. Our behavior is not purely a product of our genes. However, to deny our biological heritage would be foolhardy. Our heritage, like that of all other organisms that live on Earth, is rife with conflict.

In considering just the narrow segment of biology involving the sex ratios of organisms, we have encountered conflict of many forms. All organisms battle to survive in the harshness of the environment to which they are exposed. To survive, organisms kill, compete with, consume, enslave, or otherwise use and abuse other organisms. The huge variety of reproductive strategies among sexual organisms is the result of conflicts in interests between males and females. Parents and their offspring are frequently in conflict with each other. Siblings compete with one another. Queen and worker bees each act to try to produce within their nest the sex ratio most in their own interests. Parasites and their hosts indulge in never ending coevolutionary arms races for ascendancy. Even within the individual there is conflict, as genes vie for transmission into future generations. Truly altruistic behavior, defined as acting to increase another individual's lifetime number of offspring at a cost to one's own survival and reproduction (Krebs and Davies 1981), can and has evolved. But such examples are a tiny drop in an ocean of antagonism. We humans may have risen a very small distance out of the seas of biological conflict in at least some of our behaviors. So we teach our children, passing knowledge down the generations, promoting cultural evolution. While it is laudable to teach our children to be kind and compassionate, to compromise rather than confront, to be magnanimous in victory and gracious in defeat, we would be failing in our parental role if we did not also teach them how to deal with conflict. The altruist, unless wary, will always lose out to the cheat, the con artist, the philanderer. So my children will continue to play hide-and-seek and musical chairs and tag. And in the summer, I will stand cheering each of them on in the egg-and-spoon, sack, and obstacle races at their school's sports day.

This book has been about aspects of sex. It has been about sex determination, sex ratios, and the evolutionary implications of the differences between males and females. Much of this book has also been about microorganisms that live in larger organisms and are passed from one generation of these hosts to the next. In this final chapter, I would like to give a personal overview of the importance of these subjects and consider whether the manipulative talents of *Wolbachia* and other inherited endosymbionts may be harnessed by humanity to its own benefit. However, I will begin this discussion by looking at a question that I left unanswered at the end of chapter 1—Why did sex evolve?

The Evolution of Sex

Why sex evolved is considered one of the most intriguing and important unanswered questions in evolutionary biology. Although I say that the question is unanswered, this is only partly true. In fact, there have been many suggested answers, but none are ubiquitously accepted. Here I want to give a brief overview of the many and varied hypotheses that have been put forward.

The predominance of sex in the natural world presents evolutionary biologists with a quandary because, as previously described (p. 24), sexual reproduction is inefficient compared to asexual reproduction. This is because males of most organisms invest little in the next generation. In asexual species, all reproducing individuals invest in their progeny, either by providing nutrients for the zygotes they produce or through parental care. In species in which equal

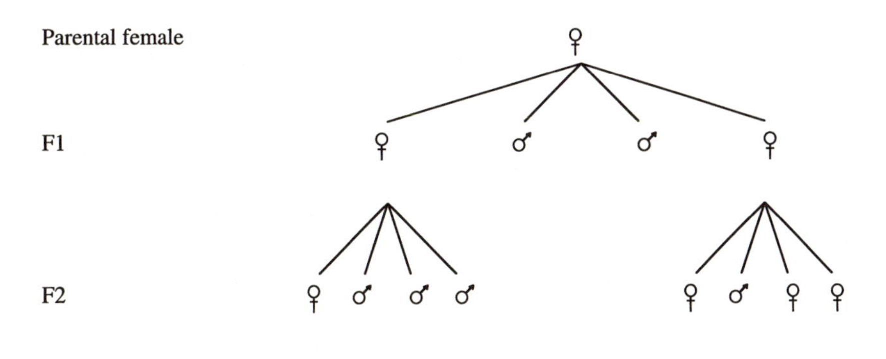

(a) Sexual foundress

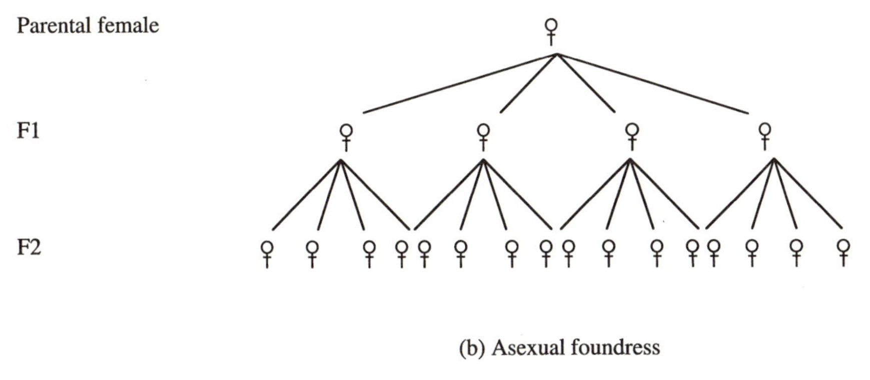

(b) Asexual foundress

Figure 9.1 Comparison of the rates of increase in a sexual species and an asexual species. The comparison assumes a 1:1 sex ratio in the sexual species and no paternal care. Each female produces four offspring.

numbers of males and females are produced, the production of males gives rise to the two-fold cost of sex (p. 24). Due to the wastefulness of males in terms of their contribution to the next generation, the production of males slows reproduction in a sexual species as compared to an asexual species. In an environment colonized by a mated sexual female insect and an equivalent asexual insect, the growth in number of the asexual form will be much more rapid than that of the sexual form (Figure 9.1).

Males are not always needless. In some species males make a significant contribution to progeny through parental care. Furthermore, in isogamous organisms, donor mating types, which may be considered as male equivalents, make the same investment in offspring as the recipient mating type. However, in most species, the female contribution to offspring far outweighs the male contribution. In addition, there are many other costs involved in sexual reproduction. These vary in different species, but a list of common energetic and/or temporal costs would include those of finding a mate, eliciting a favorable response from her/him, protecting her/him from competitors, and the act of copulation itself. There is also a risk that during close contact in courtship or during copulation one partner may contract disease from the other.

Sexual reproduction is thus costly compared to asexual reproduction, yet it has evolved frequently. According to Bell (1982), explaining the evolution of sex is "the Queen of all evolutionary problems."

The various explanations of the evolution of sex fall into two distinct classes. Either sex is seen to have evolved as the result of advantages that accrue to the individuals that indulge in it. These are termed *immediate benefit hypotheses.* Alternatively, sexual reproduction provides advantages in the longer term, through the increased genetic variation it generates. The hypotheses in this second group are called *variation and selection hypotheses* (Felsenstein 1974, 1985).

Immediate Benefit Hypotheses

Sexual reproduction may confer advantages on those that indulge in it in a variety of ways. Because in many species, reproductively mature individuals choose their mates with care, sex may increase the genetic fitness of offspring thus produced. There is now considerable evidence that females of many species choose males that carry "good genes." Males with deleterious genes will thus tend to be avoided to the detriment and possible loss of such genes from the population. This type of benefit of sex through choice of mates is likely to extend to cryptic female choice. One further immediate benefit may be that it is better to have two parents that care for offspring than one, the second parent, usually the male, being, in effect, an insurance policy against the death of the other parent.

Most mutations that have an effect on fitness are deleterious when they arise. This being the case, a further suggested immediate benefit of sex concerns the DNA repair systems that operate during meiosis (Bernstein et al. 1985, 1988). It is argued that such repair systems will mediate the number of novel mutations that are passed on to the next generation. However, Austin Burt (2000) argues convincingly against this suggestion on the grounds of the differences in the repair systems that operate in mitosis and meiosis. In mitosis, damaged chromatids are usually repaired using a sister chromatid template. In meiosis, repair is normally based on a template derived from the damaged chromosome's homolog, not its chromatid (Schwacha and Kleckner 1994, 1997). Using homologous chromosomes as the template for repair only makes sense if the function of crossing over during the first meiotic division is to generate recombinant chromosomes. This pattern thus supports hypotheses that have their origins in the suggestion, first advocated by August Weismann in 1889, that the function of sex is to generate variation for selection to act upon.

Variation and Selection Hypotheses

The variation and selection hypotheses for the evolution of sex depend on the premise that sex increases the genetic variation among progeny. In fact, sex does not create novel alleles. However, it will help to bring different alleles together into the same individual. Thus, it is perhaps more correct to say that sex generates a greater diversity of genetic combinations. Sexual reproduction produces a wide genetic array of offspring, with the genes of the parents coming together in different combinations, whereas the progeny produced by an asexual individual are in most cases genetically identical. The production of different gene combinations (Figure 9.2) may not only increase the speed with which beneficial genes can be brought together (p. 31), it may also help to break apart associations; for example, when favorable genes are associated in the same genotype with deleterious genes.

Hedging One's Bets on "Entangled" Banks

No environment is truly homogeneous. All environments that members of a species inhabit vary in both time and space. This environmental heterogeneity, exemplified in the description of a tangle of vegetation on a bank (Darwin 1859) and the enormous range of different niches such a bank would contain, may favor sexually reproducing lineages over asexual lineages. This is because, as sexually reproducing organisms produce a greater genetic variety of offspring than do asexual species, they are more likely to produce at least some progeny that are well suited to the environmental conditions that they are faced with. Perhaps a useful analogy is two people playing roulette. One puts a small stake on each of 12 numbers. The other puts a 12-times-larger stake on just one number. The first is analogous to a sexual organism, covering a range of options, while the second is like an asexual organism putting all its genetic eggs in one basket. If the wheel is a "no-house-cut" wheel—that is, it has 36 numbers and no zero—the bet-hedging gambler will win 12 times as often as the all-or-nothing gambler, but the cautious gambler wins only one-12th the amount for each win. Overall, the two strategies should have the same probability of success, but the sex or bet-hedging strategist is likely to stay in the game longer if the wheel is running for the house. In real organisms, high levels of competition, or an unusually long run of adverse environmental conditions, may lead to the extinction of an asexual species before the conditions to which its particular genotype are well suited come up.

This idea can be extended to species that show very high degrees of promiscuity. For example, female *A. bipunctata* mate much more often than is necessary, purely to maintain full egg fertility. *Adalia bipunctata* larvae and adults feed on aphids. These aphids are a very unpredictable food source, with different species that specialize on different host plants and trees doing well in different years. Each host plant will present different microclimatic conditions. Thus, the conditions faced by the progeny of a reproducing female are not predictable. One suggested reason for the promiscuity of females is that by mating

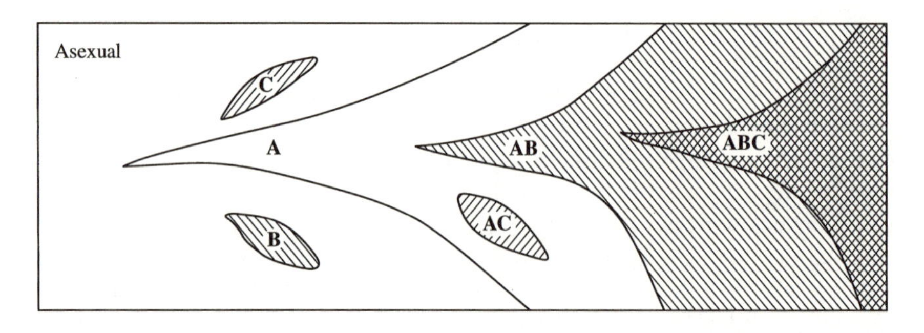

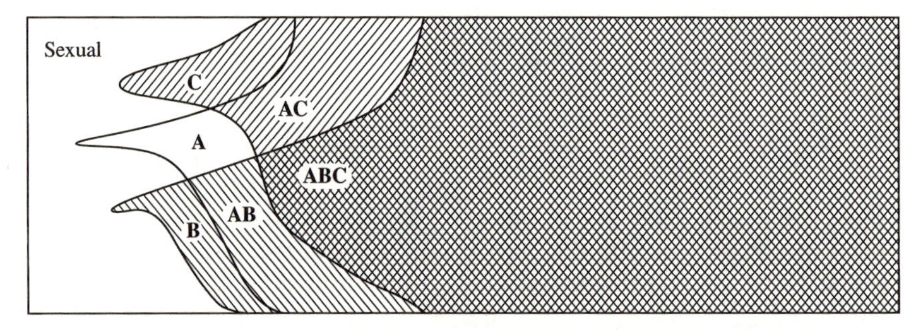

Figure 9.2 The spread of three novel advantageous mutations contrasted in asexual and sexual populations. Because of sex and recombination, individuals with all three mutations arise more rapidly in a sexual than an asexual population. (After Fisher 1930; Muller 1932.)

with a sequence of different males, the female is ensuring maximum genetic diversity among her progeny, so that at least a few have suitable traits for the conditions they face. Here then is a theory that not only may be of significance in promoting sexual reproduction over asexual reproduction, but also may promote increasing amounts of sex.

The increased variation between progeny produced as a result of sex may also help decrease the amount of competition between siblings (Bell 1982). If kin compete with one another, the variable genotypes in the progeny of a sexual female may allow for a wider range of niches to be occupied by these offspring than by those of an asexual mother, which may all be fit for the same niche.

The Red Queen

Coevolutionary arms races, such as those between predators and prey or hosts and parasites, inherently impose selection pressures on both members of an interaction. Prey or host individuals with genes that reduce the efficiency of the predator or parasite will have increased reproductive success. Therefore, genes of this type spread. These antipredator or parasite-resistance genes then impose a selection pressure on the predator or parasite to evolve countermeasures. If this selection results in successful countermeasures in the predators or parasites, novel selection pressures are imposed on the prey or hosts. Coevolutionary cycles result. These are often referred to as the Red Queen effect, deriving from Lewis Carroll's *Through the Looking Glass*. As the Queen remarks:

> "Now, here, you see, it takes all the running you can do, to stay in the same place. If you want to get somewhere else, you must run at least twice as fast as that."—Lewis Carroll, *Through the Looking Glass*

In the same manner, in evolutionary terms a prey or host species may have to run, not to improve, but to avoid being overwhelmed. Similarly, a parasite must run to avoid being outwitted by its host.

We have already seen that sex increases the rate of evolution. It then follows that for species in highly competitive situations, faced not only with competition for resources but also with a range of predators and parasites, sexual lineages will do better than asexual ones. In the case of disease parasites, as Hamilton et al. (1990) have proposed, males, although a costly invention, may have the saving grace that they are necessary for health.

How important the advantage of sex is in coevolutionary arms races, and whether the need to counteract predators and parasites is an important force in the evolution and maintenance of sex, awaits appropriate empirical evidence (Burt 2000). Obtaining such evidence may be difficult because, in many organisms, sex and disease are not independent. One only has to think of the wide range of diseases that are transmitted during sexual contact to realize that this is so. Thus, sex may create a problem of parasites as well as solving it.

Deleterious Mutations and the Evolution of Sex

It is arguable that the most appropriate genotype for a given environment can never be maintained indefinitely, even if the environment remains constant, because mutation is a ubiquitous property of genetic material. Even in asexual populations, variations between lineages are created through mutation. Because deleterious mutations are commoner than advantageous mutations, it is fair to say that, on the whole, mutation tends to destroy adaptation. Because asexual organisms cannot purge deleterious mutations by recombination, strongly detrimental mutations will be lost from a population by selection within a generation or two. However, slightly deleterious mutations will accumulate in populations as a result of Muller's Ratchet (see p. 32). This places a cost on being asexual. To illustrate this process, Ridley (1993) imagines asexual reproduction to be akin to a photocopying process. Modern photocopiers produce excellent, but not perfect, copies of documents. If each new copy is made not from the original, but from the previously made copy, slight blemishes will gradually build up in each copy "generation." The quality of copies thus decreases in each generation, in the same way that the fitness of asexually reproducing organisms decreases as blemishes or mutations build up. Sexual organisms do not accumulate blemishes in the same way because the best sections of different copies within each generation of copies can be combined to produce a new generation that is as good as the original.

A reduction in the mean fitness of individuals in a population does not depend on the loss of the best type through the ratchet process. The recurrent generation of deleterious mutations will lower population fitness as a matter of course as long as more mutations are deleterious than advantageous. As mentioned in chapter 1, Kondrasov (1982, 1984) has suggested that sex will act to decrease the contamination of the genome by deleterious mutations because sex increases variation in the individuals in a population with respect to the number of harmful mutations they carry. In a sexual population, some individuals will have many harmful mutations, but some will have very few. In an asexual population, the variance in the number of deleterious mutations between individuals will be lower. If selection removes the least fit 25% of the population in both sexual and asexual populations, more deleterious mutations will be removed from the former than from the latter (p. 33).

The Enigma of Sex

The evolution of sex remains controversial. Each of the different theories has its advocates and detractors. For each of the theories, there are counterarguments (e.g., see Barton and Charlesworth 1998; West et al. 1999). The evolution of sex may be similar to evolution as a whole. The polarity of the selection versus random genetic drift controversies of the 20th century were resolved toward the end of the century into a synthesis in which both neutral drift and natural selection were seen to be important evolutionary processes. Similarly, it is feasible that each of the ideas for the evolution of sex plays its part in some instances and that, in many cases, two or more factors have combined to give sexual lineages an advantage over asexuals, either in the short term or the long term. Not only could different factors favoring sex combine, but mutational and environmental factors could interact synergistically in many ways. A pluralistic approach to the explanation of sex is probably the most likely route to a solution for this "queen of problems in evolutionary biology."

Testing the alternative hypotheses for the evolution of sex is not going to be easy. One problem is the difficulty in distinguishing those factors that were important in the initial evolution of a sexual lineage within an asexual population from those factors that maintain sexual reproduction once it has evolved. Here we have a similar difficulty to that encountered by researchers considering the evolution and maintenance of sexual selection through female choice. While many models seem to effectively cause the evolution and spread of elaborate secondary sexual characteristics in males once a significant proportion of females are making a choice, the evolution and initial spread of female choice genes is a very slow process.

Sex is a complex and delicate process. The central process of sexual reproduction is the production of gametes by meiosis. This involves the accurate copying of genetic material, the recombination of this material, and the precise division of homologous chromosomes as cells divide, so that the gametes produced each contain the correct number of chromosomes without excessive loss of genes through inaccurate recombination. This must be followed by the successful fusion of appropriate gametic cells. The theories outlined above may give some indication about why sexual reproduction is so common compared to asexual reproduction in the natural world, and may help point to the types of selection that were important in the initial evolution of sex. However, they do not explain the mechanistic pathway that allows accurate meiotic sex. Despite the huge recent advances in cell biology and molecular and developmental genetics, this problem, perhaps unsurprisingly, remains unresolved.

The Importance of Ultraselfish Genes

I now turn to the question of the importance of ultraselfish genes in biology. Ultraselfish genes, genes that spread through host populations despite the harm that they inflict on their hosts, together with the genes that evolve to suppress or mediate their effects, provide evidence of conflicts between different parts of the genome. One important aspect of these genes is not in dispute. They show that the fundamental unit of selection is the gene (or linkage group). The extreme spite of driving sex chromosomes and cytoplasmic sex ratio distorters, in addition to transposable elements, driving B chromosomes and autosomes, have been cited again and again as extreme examples, showing that the phenotype is purely a gene carrier.

For a long time, ultraselfish genes were widely viewed, by those aware of their existence, as genetic oddities—exceptions that proved certain rules but had little other significance in evolutionary biology. I hope that in this book I have shown that this is not the case.

J.B.S. Haldane first suggested, as early as 1932, that ultraselfish genetic elements may have considerable evolutionary importance. However, this contention was not considered seriously until Hamilton (1967) argued that such elements were critical in understanding some examples of deviations in the sex ratios of organisms from the parity of males and females seen in most species. Cosmides and Tooby (1981), with an analysis of the evolutionary interests of cytoplasmic genes in contrast to nuclear genes, and then Werren et al. (1988), in a review of the types of ultraselfish genes, heightened interest. From that time, interest in ultraselfish genes has had a meteoric rise. The intragenomic conflict manifest in the existence of ultraselfish genes is now recognized to have been and be influential in a great range of evolutionary genetic processes, and indeed may have played a crucial role in the evolution and design of many genetic systems. Papers theorizing and, in most cases, demonstrating a role for intragenomic conflict in the evolution of sexual reproduction, the evolution of the sexes, the evolution of sex-determining mechanisms, the evolution of sex ratios, the evolution of sex allocation systems, the evolution of mating strategies, and in speciation, have all been published in the last decade.

Potential intragenomic conflict is ever-present in all organisms. Its presence imposes selective pressure on genetic systems to keep the anarchy it produces in check, so that outbreaks of genomic warfare are rare. However, when they do occur, they have a major impact on evolution. Just as human technology, behavior, and morality tend to change more during periods of war than when peace pervades, so genome wars lead to rapid evolutionary response as a result of the strong and far-reaching selection pressures imposed on the individual by different parts of its own genome.

Much of the recent work on intragenomic conflicts has centered on cytoplasmically inherited symbionts (Werren and O'Neill 1997). This is because these are exclusively maternally inherited and so are obviously in conflict with genes that are inherited from both parents. Many of these inherited symbionts affect the sex ratios of their hosts, favoring females. Investigations into the ways in which the sex ratios of sexual species are manipulated to their own ends—either by genes of the organism or by what O'Neill et al. (1997) termed *influential passengers*, inherited microbes within the organism—are still in their infancy. Although scientists have known about extraordinary sex ratios in a variety of invertebrate taxa for a considerable time, most of the techniques that allow their close scrutiny are relatively recent developments. Novel techniques and advances in the study of the agents that cause sex ratio distortion are now regularly being published in scientific journals. Most effort is currently proceeding in three areas: how widespread and diverse are sex ratio distorters among invertebrates; what are the mechanisms of the various types of sex ratio distortion; and whether and how the agents of sex ratio distortion can be harnessed as tools for humanity.

A fourth question, what part sex ratio distorters have had in the evolution of their carriers, has received less attention. Trying to work out the past effects that sex ratio distorters may have had on their hosts is by its very nature a rather hit-and-miss process. Yet advances are being made in this area as well. Many of these advances are theoretical; for example, the work of Johnstone and Hurst (1996) on the effect that invasion by sex ratio–distorting bacteria has on the mitochondrial DNA variation of a host population. This type of work is of particular value because it makes testable predictions, and the predictions can largely be tested in the laboratory using molecular genetic techniques.

More difficult to address are the evolutionary consequences that biases in population sex ratios may have on the reproductive biology of their hosts. Some predictions can, of course, be made. For example, a sex ratio bias in a monogamous species is likely to have a greater evolutionary influence than a similar bias in a polygamous species. But how strong the bias would have to be in a polygamous species before there would be an impact is more difficult to ascertain. For example, in *Acraea encedon* and *Acraea encedana* the high prevalence of male-killing *Wolbachia* in some populations of these species has affected the mating behavior, males becoming the limiting sex as a result of their rarity compared to females. But should we expect the same effect in *Harmonia axyridis* or *Adalia bipunctata*, where the sex ratio biases due to male-killing are less extreme and the reproductive ability of males—that is, the number of females that a single male can service—is far greater?

Is Wolbachia *Special?*

Throughout this book, one genus of bacteria, *Wolbachia*, has cropped up with considerable regularity. The ability of *Wolbachia* to play different games with their hosts is fascinating. They are mutualistic in some hosts, such as some of the human disease-causing species of nematode. In other hosts they impose a cost through the manipulation of their hosts' reproductive systems, and appear able to switch strategies depending on the ecological and genetic constraints imposed by their hosts and under which they are forced to operate. Their abilities seem extraordinary. We should ask, therefore, whether *Wolbachia* is indeed out of the ordinary, or whether other groups of microorganisms can play a variety of roles in their hosts. This is not an easy question to answer, for there is certainly some study bias resulting from the molecular methodologies now used to hunt for intracellular bacteria. We must therefore ask this: Does *Wolbachia* employ a greater variety of strategies because it alone has the ability to employ all these strategies, or does it simply appear to do this because it has been more sought after and closely studied than other groups of microorganism?

To answer this question, we need to look at only those studies that were host phenotype–led. That is to say, we need to look at those case histories that began with an observation of an abnormality in an invertebrate's reproductive system, or sex ratio. We must ignore those cases in which the presence of *Wolbachia* was ascertained before the effect it had on the host was discovered. Doing this, we find CI to be specifically the province of *Wolbachia* or other members of its family, the Rickettsiaceae. Parthenogenesis induction by microbes also appears to be exclusively the province of *Wolbachia*. On the other hand, feminization and male-killing of hosts are employed by a wider range of microbes, and, at least in the latter, it is certain that the full list of bacteria that employ these strategies has yet to be completed. This suggests that *Wolbachia* are indeed exceptional, for they are the only group that employ all four manipulative strategies. Interestingly, CI and PI, which appear to be almost exclusively the province of *Wolbachia*, have something in common. In both, the mechanisms by which the bacteria affect their hosts involve manipulation of chromosomes. It is possible that only *Wolbachia* has evolved this mechanism. In the case of feminization and male-killing, the mechanics of the manipulations are not known, although French scientists are probably close to discovering how *Wolbachia* feminizes some crustaceans. It is possible that the diversity of feminization and male-killing agents reflects the fact that a variety of different pathways of changing the sex of genetic males or recognizing and differentially killing male hosts exist. If this is the case, different microbes may employ different techniques in these instances.

The tools for studying endosymbionts are advancing rapidly. The various techniques of labeling specific individuals with markers attached to specific genes should allow questions of the mechanisms of genes and bacteria that distort invertebrate sex ratios to be far more easily addressed. For example, the development of an amplification system for the "glow" gene from the jellyfish *Aequorea victoria* allows us to insert a noninvasive heritable marker in endosymbiont hosts. This raises tremendously exciting possibilities for efficient experimentation on these hosts, since determination of genotypes will not have to be assessed by killing hosts and subjecting them to molecular analysis post hoc.

Similarly, efficient techniques for reversibly knocking out the expression of precisely specified genes, without affecting the expression of any other genes, should allow a wide range of questions relating to the genetic interactions of endosymbionts and their hosts to be addressed. Two questions at the top of the list would be whether the same gene or genes are involved in the various host manipulations employed by *Wolbachia* and whether the diverse types of bacteria that practice male-killing have male-killing gene(s) in common.

Novel and powerful molecular genetic tools are being widely applied to ultraselfish symbiotic bacteria. Currently, *Wolbachia* are in fashion. In 2000 over 100 scientists from research groups all over the world met in Crete for the First International Conference on *Wolbachia*. The range of topics—from the manipulative strategies of parasitic *Wolbachia* through their population dynamics and impacts on the evolution of their hosts to their importance in human diseases caused by nematodes such as the infamous eye worm, Loa loa, and elephantiasis (Plates 8c and 8d)—was enormous. In particular, the revelation of two multimillion-dollar projects to sequence the entire genomes of a variety of strains of *Wolbachia* gave evidence of the import being attached to this genus of bacteria (O'Neill et al. 2000; Bourtzis et al. 2000).

Implications of Sex Ratio Distorters for Evolutionary Geneticists

It is becoming increasingly clear that inherited symbionts occur very commonly in invertebrates. Many of these symbionts are beneficial. However, a sizable proportion are deleterious and, rather than being beneficial to their hosts, they are antagonistic toward them, or at least toward some of them. The existence and common occurrence of such symbionts has several important consequences. First, at the most general level, it is difficult to conceive that a thorough understanding of the overall biology of any arthropod species can be achieved without an appreciation of the symbionts it carries, whatever the nature of the interactions between the host and its symbionts. Coevolutionary pressures, whatever their direction, are likely to have considerable impact on both members of an association. In the case of deleterious symbionts, it is already apparent that symbionts may fundamentally alter host sex determination, may reverse the direction of sexual selection, or may cause the evolution of symbiont suppressor and rescue genes. Other evolutionary consequences of the invasion of deleterious symbionts surely await discovery.

Second, at a more practical level, the spread of any symbiont through a host population will cause an immediate decrease in the diversity of mitochondrial

DNA (mtDNA) in that population. Because both symbiont and mitochondria are cytoplasmically inherited, they are effectively linked to one another. The spread of a symbiont from one, or a small number of, initially infected host females, therefore, pulls the mitotypes of these founders through the population. Effects will be seen for all deleterious symbionts, although the effects in the case of male-killers may be greater than for other systems, due to the usually low prevalence of male-killers and the constant recruitment of uninfecteds from infecteds because of imperfect vertical transmission. It is thought unlikely that selective sweeps imposed on mtDNA by the spread of symbionts will impact significantly on the adaptive evolution of hosts. However, for evolutionary biologists who use mtDNA as a tool to study gene flow, population size, or population history, the effects of symbiont spread on mtDNA diversity mean that in such species, mtDNA cannot be used as a neutral marker. Lack of mtDNA diversity in a population need not reflect low effective population size or genetic bottlenecks. It may simply reflect past or current presence of an inherited symbiont.

Can Sex Ratio Distorters Be Useful to Humanity?

The full title of the recently funded European *Wolbachia* Genome Project is *The European Wolbachia Project: Towards Novel Biotechnological Approaches for Control of Arthropod Pests and Modification of Beneficial Arthropod Species by Endosymbiotic Bacteria* (Bourtzis et al. 2000). This title highlights the fact that scientists envision potential uses of ultraselfish endosymbionts both as control agents of insect and other invertebrate pests and as aids for arthropods that are beneficial to man. Let us first consider ways in which beneficial invertebrates may be helped.

Helping Beneficial Insects

Several strategies to aid beneficial invertebrates can be envisioned. First, endosymbionts that bias biological control agents, such as parasitoid wasps or aphidophagous coccinellids, in favor of females, may be used to increase the impact of the control agent, or reduce costs of a control program. In the case of parasitoid Hymenoptera, only females have a negative effect on the host. Thus, a heritable symbiont that caused only females to be produced would be commercially beneficial.

Benefits may also arise from male-killers in coccinellids if the presence of the symbiont increased the density of females in the population, even if this increase were essentially at the expense of males. This is because the food intake of female coccinellids throughout their lives is significantly greater than that of males.

One interesting case that is currently being put into practice involves improving the commercial efficiency of ladybirds as biological controllers of aphids using a naturally occurring male-killing *Spiroplasma*. Ladybirds are major predators of aphids, but have a poor record in the field as controllers of these sap-sucking plant pests for three reasons. First, ladybirds are expensive to produce in mass cultures using aphids as food, thus making them uncompetitive compared with other forms of control. Second, ladybirds tend to fly away when aphid densities are below a certain threshold level; as a consequence, farmers are reluctant to use them because of their unreliability. Third, the return on initial investment for the supplier may be limited if the released control species establishes itself in the wild and reduces the need for control in the future. While I would hope that most people would see this last possibility as desirable, it will not encourage financial investment in biological control from independent companies.

André Ferran's research group in Antibes are researching the Asian ladybird *Harmonia axyridis*. Their intent is to develop this species as an aphid control agent for use in Europe. In their work, they have gone a considerable way toward solving some of the problems that ladybirds present. First, they have found an alternative diet for the ladybirds, the eggs of the moth *Ephestia kuehniella*, which, although expensive, is cheaper than using aphids, since the man-hours involved in culturing is considerably reduced. Second, they have selected mutant strains with reduced wing muscles that make flight impossible. The plan now is to introduce the male-killing *Spiroplasma* that occurs naturally in this species in Asia into the flightless stocks. This will serve two purposes. First, it will allow the culture populations to be maintained with a strong, managed female bias, making captive breeding programs more economical. Second, if managed carefully, the coccinellids released into the wild can all be mated females that are infected with the male-killer. This will mean that the species will not be able to establish itself in the wild, for although the females will produce progeny where released, all the progeny will be female. The problems of releases of nonnative biological control agents that have been experienced in many parts of the world in the past should thus be avoided.

Wolbachia *and Pest Control*

Ultraselfish symbionts may prove to be of use in improving the value of beneficial insects to man. They have even more potential in pest control (Siskins et al. 1997). Most useful of the ultraselfish genetic elements in this regard seem to be the CI *Wolbachia*. This is because these bacteria tend to spread through novel populations to fixation. Thus, one strategy would be to employ CI agents, by mass release, to cause sterility in recipient populations, much in the way that mass releases of males sterilized by radiation were used to control *Cochliomyia hominivorax* in several countries (e.g., Krasfur et al. 1987). Such a methodology has already been used in pilot trials with a variety of targets, with varying degrees of success (Siskins et al. 1997).

An alternative would be to employ symbionts as gene transport vectors. The basis of this idea is that if a novel symbiont bearing "useful" genes were likely to spread to high prevalence once introduced into a target population, as might be expected of a CI agent, the useful genes would be carried through the population from a relatively small initial release. A useful gene in this context would have to fulfill two critical requirements. First, it would have to be transferable into an inherited symbiont and be vertically transmitted thereafter with almost perfect efficiency. Second, it would need to be expressed on the genetic background of the pest species; that is, it would have to be a dominant gene. Genes fulfilling these requirements and occurring naturally in inherited symbionts are improbable. Therefore, if this idea is to become a reality, it will require the use of recombinant DNA technology to produce transgenic insect strains. If appropriate genes can be found and transferred into appropriate hosts, releases of transgenic strains that will spread through host populations as a result of the inherent population dynamics of ultraselfish endosymbionts show great promise. Certainly their potential as a genetic system to modify field populations compares favorably with drive methodologies of nuclear genes.

Obviously, the implications of such programs must be submitted to the most stringent theoretical and empirical testing. These tests need to evaluate the economic viability of the various systems that come to the fore. More important, the potential dangers of release of transgenic endosymbiotic strains also need to be submitted to the most rigorous analysis. Furthermore, theoretical analyses must then be shown, by empirical evidence, to be applicable and pertinent to each and every system developed before it is released. Computer simulations alone must not be seen as sufficient. One has only to think of the multiple male-killers of *A. bipunctata* in Moscow, which are in direct conflict with current theory, to see that nature does not always agree with evolutionary theorists.

That said, I think that there is absolutely no doubt that *Wolbachia* and other inherited microorganisms will receive greatly increased attention over the next decade or two because they have the potential to be harnessed by humankind.

Six Impossible Things

> "Why, sometimes I've believed as many as six impossible things before breakfast."—Lewis Carroll, *Through the Looking Glass*

Charles Darwin (1887) wrote in his autobiography:

> I have steadily endeavoured to keep my mind free, so as to give up any hypothesis, however much beloved (and I cannot resist forming one on every subject), as soon as facts are shown to be opposed to it.

Such an attitude is simply good science. Sadly, it is an attitude that does not pervade the whole scientific community. Those who publicize speculative theories that are then proved incorrect are frequently criticized to the detriment of their careers and credentials. This means that many biologists are reluctant to cast forth the fruits of their imaginations. Given the poor, sensationalist, and frequently distorted reporting of science in our "sound-bite" society, few scientists would not have sympathy for those who act cautiously with their ideas. However, there can be no doubt that such caution can only be to the detriment of scientific discovery and advance. As British Telecom has noted in many of its advertisements, we do so much better if we "just keep talking."

I am foremost a naturalist and, like Darwin, I cannot help speculating. I also have a tenured position so I have the security to speculate in public. This book has been about sex, sex ratios, and sex ratio distorters, primarily in the invertebrates. I will close with a small amount of speculation, by making six predictions about sex ratios in invertebrates. I have no idea whether these predictions will be fulfilled or not, so it may be that, like Alice, I have believed as many as six impossible things before breakfast. For what they are worth, here they are, and I hope that some biologist somewhere is prepared to put them to the test.

1. Males in some populations in which the sex ratio is strongly female biased due to the action of male-killers and in which males are promiscuous will pass less sperm in each copulation than in less female-biased populations.

2. Over 80% of ladybird beetles that feed on aphids and lay eggs in clutches will be found to harbor male-killing bacteria, or carry male-killer suppressor or male rescue gene(s) that evolved in response to past invasion by male-killing microbes.

3. Instances of *Wolbachia*, or possibly some other inherited microorganism, that adversely affect the male reproductive parts or sperm production in hermaprodite invertebrates will be found.

4. Inherited bacteria other than *Wolbachia* will be discovered that both cause feminization and induce parthenogenesis.

5. Ultraselfish behaviors will be discovered in every major group of inherited bacteria of invertebrates.

6. Male-killing will be found to be a main factor in the evolution of color pattern genetic polymorphism (*sensu* Ford 1940) in aposematic species of butterflies, ladybirds, and other beetles.

GLOSSARY

Adaptation — Any characteristic of an organism that enables it to cope better with conditions in its environment.

Adenosine triphosphate (ATP) — An energy-rich molecule that promotes many cellular activities.

Aestivation — Summer dormancy, when conditions become unfavorable.

Allele — When a gene or other DNA sequence exhibits variability, each alternative form is known as an allele.

Altruistic — Any behavior by which an individual increases the fitness of a conspecific, without gaining a fitness benefit itself.

Amphitoky — The form of parthenogenesis in which unfertilized eggs develop into either sex. Also referred to as *deuterotoky.*

Anisogamy — Anisogamous species are those in which the male and female gametes are of different sizes.

Apomixis — A form of parthenogenesis whereby neither chromosome reduction nor fusion of nuclei nor any corresponding phenomenon takes place in the egg.

Aposematic coloration — Conspicuous color patterns associated with distasteful, toxic, or otherwise harmful characteristics of an animal.

Arms race — A coevolutionary pattern involving two or more interacting species (e.g., predator and prey, or parasite and host), where evolutionary change in one party leads to the evolution of a counteradaptation in the other.

Arrhenotoky — The form of parthenogenesis in which unfertilized eggs develop into males and fertilized eggs develop into females.

Arthropod — Member of the phylum Arthropoda. Animals with jointed bodies and limbs, and an exoskeleton.

Asexual reproduction — Any form of reproduction that does not involve the formation and union of gametes from the different sexes or mating types.

Assortative mating — Mating in which partners are chosen because they are phenotypically similar with respect to one or more phenotypic characteristics.

Automitosis — A form of parthenogenesis whereby regular chromosome pairing and reduction of chromosome number occur during gamete formation, but in which chromosome number is restored by the fusion of two haploid nuclei.

Autosome — A chromosome other than the sex chromosomes.

Balanced polymorphism — A genetic polymorphism in which the various forms are maintained in the population at constant frequencies, or their frequencies cycle in a more or less regular manner, and both alleles are maintained above a frequency that could be maintained by recurrent mutation.

Barr body sex test — A test of the sex of an individual of a placental mammal dependent on the presence of a condensed mass of chromatin (the Barr body) in nuclei that contain more than one X chromosome.

B chromosome — Any chromosome occurring in animals or plants that differs from normal chromosomes (autosomes and sex chromosomes) with respect to morphology, number, meiotic behavior, and/or genetic effectiveness.

Bilateral mosaic — An individual that appears split down the middle, having different features on one side compared to the other. Most are individuals that appear half-male and half-female.

Biparental inheritance — Genes and genetic elements in sexual organisms that are inherited from both parents (e.g., most nuclear genes) (see **Uniparental inheritance**).

Bursa copulatrix — The region of the female genitalia of a invertebrate that receives the male intromittent organ and sperm during copulation. Its structure is often important in distinguishing closely related species.

Centromere — The region on a chromosome that becomes attached to the spindle during cell division.

Chiasma (plural: *chiasmata*) — Regions of contact between homologous chromosomes during pairing at meiosis (from late prophase to the beginning of anaphase I) at which crossing over takes place to allow the exchange of homologous sections of nonsister chromatids.

Chloroplast — An organelle in the cytoplasm of plants that contains the green pigment chlorophyll and in which starch is synthesized. The main site of photosynthesis.

Chromatid — In cell division, one of the two identical strands resulting from self-duplication of a chromosome.

Chromosomes — Small, elongated bodies, consisting largely of DNA and protein, in the nuclei of most cells, existing generally in a definite number of pairs for each species and generally accepted to be the carriers of most hereditary qualities.

Clade — A single whole branch of a phylogeny.

Cline — A gradual change, within an interbreeding population, in the frequencies of different genotypes, or phenotypes, of a species.

Coevolution — Occurrence of genetically determined adaptive traits in two or more species as a result of selection due to mutual interactions controlled by these traits.

Competition — The interaction of individuals or species resulting from the use of limited resources.

Competitive exclusion principle — The principle that ecologically identical species cannot exist in the same habitat for an indefinite period.

Condition-dependent handicaps — Secondary sexual traits that are potentially disadvantageous to their bearers, due to reduced survival chances, but that may confer a reproductive advantage by being the subject of mating preference. The development of these traits is positively correlated to the condition of the individual bearing them; therefore, they have the potential to act as fitness indicators.

Conjugation — Union of the sex cells or unicellular organisms during fertilization.

Conspecific — Of the same species.

Crossing over — An interchange of groups of genes between the members of a homologous pair of chromosomes.

Cryptic coloration — When the colors of an animal match those of the environment in which it lives.

Cryptic female choice — Female choice occurring through a variety of mechanisms after copulation that affect which sperm succeed in fertilizing eggs.

Cytoplasm — The living substance of the cell, excluding the organelles within it.

Cytoplasmic genome — The entire genetic complement of a cell that lies outside the nucleus of the cell. It includes the genetic composition of organelles that lie within the cytoplasm and any symbionts that live within the cytoplasm.

Cytoplasmic inheritance — Hereditary transmission dependent on the cytoplasm or structures in the cytoplasm rather than in the nucleus.

Cytoplasmic male sterility (CMS) — The phenomenon in some higher plants whereby a maternally inherited factor prevents the production of viable pollen.

Deoxyribonucleic acid (DNA) — The principal heritable material of all cells. Chemically it is a polymer of nucleotides, each nucleotide subunit consisting of the pentose sugar, 2-deoxy-D-ribose, phosphoric acid, and one of the five nitrogenous bases adenine, cytosine, guanine, thymine, or uracil.

Diapause — Temporary interruption in the normal reproduction or development of an organism. Usually associated with a dormant period when the habitat is unfavorable.

Dioecious — Of plants in which the reproductive organs of the male and female sex occur on different individuals.

Diploid — An organism or cell with two sets of chromosomes or two genomes.

Direct selection — In sexual selection, when the mating choice made by an individual (usually the female) leads to an increase in the number of progeny it raises during its lifetime.

Directional selection — Selection in which individuals at one end of the distribution of a trait showing continuous variation are at an advantage over other individuals.

Disassortative mating — A preference to mate with individuals having a different genotype to one's own.

Disruptive selection — Selection in which individuals at both extremes of the distribution of a trait showing continuous variation are at an advantage over intermediate individuals.

DNA — *See* **Deoxyribonucleic acid.**

DNA fingerprinting — The first genetic method capable of identifying individuals uniquely, based on minisatellite DNA sequences and invented by Professor Alec Jeffreys. Often used more loosely to refer to any genetic method that identifies individuals with high confidence.

Dominant — The opposite of recessive. The stronger of a pair of alleles, expressed as fully when present in a single dose as it is when present in a double dose.

Dorsal — Concerning the back or upperside of an animal.

Dosage compensation — The mechanism by which the activity of a gene is increased or decreased according to the number of copies of that gene in the cell.

Ecdysis — The molting process, by which an insect changes its outer skin.

Eclosion — The metamorphosis of a pupa into an adult insect.

Effective population size — The effective size of the population is the size of an ideal population that would exhibit equivalent properties with respect to genetic drift.

Electrophoresis — A technique used to detect differences in proteins and polypeptide chains based on differences in electrical charge of the molecules.

Elytra — The hard and horny front wings of a beetle (singular elytron).

Endomitosis — Duplication of chromosomes without cell or nuclear division in the normal sense. In some parthenogenetic organisms, diploid zygotes are produced by endomitosis.

Endoparasite — A parasite that lives inside the body of its host.

Endosymbiont — An organism that lives within the cell or body of another.

Enzyme — A protein that initiates or facilitates a chemical reaction between other substances.

Eukaryotes — Single or multicellular organisms in which the cell(s) contain a well-defined nucleus, with multiple linear chromosomes, which is enclosed by a nuclear membrane.

Evolution — A cumulative, inheritable change in a population.

Evolutionary stable strategy (ESS) — A strategy such that, if all members of a population adopt it, no alternative strategy can invade.

Exoskeleton — The external skeleton of an insect; made of cuticle.

Extracellular — Outside the cell.

Fecundity — The number of eggs or offspring produce by an organism.

Female choice — One of Darwin's two mechanisms of sexual selection. The phenomenon whereby females exercise choice over which of two or more available males they mate with.

Fertilization — The fusion of a male gamete and a female gamete to form a zygote.

Fisherian mechanism of sexual selection — The spread of a female-choice gene because of genetic correlation with the preferred male trait, the spread initially owing to a naturally selected advantage, and later being associated with a sexually selected advantage.

Fisherian runaway — The theoretical outcome of Fisher's mechanism of the evolution of female choice, involving a selective feedback loop between a female preference for a male trait and the trait itself. Used as an explanation of extreme male ornamentation, the process runs until arrested by the fitness costs of the extreme trait.

Fitness — The relative ability of an organism to transmit its genes to the next generation.

Fixation — Situation in which an allele or genetic element has spread to all members of the population that carries it.

Founder effect — Any change in the genetic constitution of a population that results entirely from the process of sampling. Such changes are strongest when the sample size is small. Founder effects are usually associated with colonizing or founding populations.

Frequency-dependent selection — Selection in which the fitness of a genetic form is dependent on the frequency of that form in the population relative to others.

Gamete — Reproductive cell.

Gametogenesis — Formation of male and female gametes or sex cells.

Gene — A hereditary factor or heritable unit that can transmit a characteristic from one generation to the next. Composed of DNA and usually situated in the thread-like chromosomes in the nucleus.

Gene flow — Genetic exchange between populations resulting from the dispersal of gametes, zygotes, or individuals.

Gene locus — The position on a chromosome at which a particular gene resides.

Genetic bottleneck — The reduction in genetic variability in a population resulting from a period of low population size.

Genetic drift — Random changes in the frequency of genes in populations.

Genetic linkage — The tendency for certain genes to be inherited together, instead of assorting independently, because they are situated close to one another on the same chromosome.

Genetic markers — Any allele (or sequence) that is used experimentally to identify a sequence, gene, chromosome, or individual.

Genetic polymorphism — The occurrence together in the same population, at the same time, of two or more discontinuous, heritable forms of a species, the rarest of which is too frequent to be maintained merely by recurrent mutation (*sensu* Ford 1940).

Genetics — The study of inheritance.

Genome — The entire complement of genetic material in a cell. In eukaryotes this word is sometimes used to refer to the material in just a single set of chromosomes.

Genotype — The genetic makeup of an individual with respect to one genetic locus, a group of loci, or even its total genetic complement.

Germ cells — Those cells destined to become the reproductive cells.

Gonad — A sex organ (ovary or testis) that produces gametes.

Gonadal ridge — Embryonic structure composed of cells that develop into the primary sex organs (testes and ovaries).

Group selection — Natural selection acting in favor of the survival of a group of individuals of a species, but not necessarily that of each individual.

Gynandromorph — An individual of a dioecious species that is a sexual mosaic, typically having some portions of the body male and other portions female.

Habitat — The specific place and type of local environment where an organism lives.

Handicap principle — In sexual selection, the principle that any trait chosen by a female as an indicator of genotypic or phenotypic quality can only be considered a good indicator if it can only be expressed by males of high quality.

Haploid — Of cells (particularly gametes) or individuals having a single set of chromosomes.

Haplo-diploid — Organisms having one sex haploid and the other diploid. Based on a sex-determining mechanism by which females develop from fertilized eggs and functional, fertile males from unfertilized eggs.

Hemolymph — The mixture of blood and other fluids in the body cavity of invertebrates.

Heritability — The proportion of the total variation in a trait that is due to genetic rather than environmental factors.

Hermaphrodite — An individual that bears some tissues that are identifiably male and others that are female, and that can produce both mature male and female gametes.

Heterogametic sex — The sex that carries sex chromosomes of different types and thus produces gametes of two types with respect to the sex chromosome they carry. In humans, males are the heterogametic sex, carrying one X and one Y chromosome.

Heterozygote — An individual that bears different alleles of a particular gene, one from each parent.

Heterozygote advantage — The situation in which the fitness of heterozygote individuals exceeds that of either type of homozygote.

Hitchhiking gene — Any gene that increases in frequency as a consequence of being associated by genetic linkage with a selectively advantageous gene, rather than as a consequence of its own fitness effect on its bearer.

Homogametic sex — The sex that carries sex chromosomes that are the same. In humans, females are the homogametic sex, carrying two X chromosomes.

Homologous — Genes, other DNA sequences, or entire chromosomes are said to be homologous if the similarities they share result from a common ancestry.

Homozygote — An individual that bears two copies of the same allele at a given gene locus.

Horizontal transmission — The transmission of a gene or genetic element from one organism to another by any mechanism other than that normally involved in the inheritance of genetic material by offspring from parents.

Hormone — Any type of chemical messenger responsible for the timing and regulation of metabolic, behavioral, or other processes.

Host — An organism that is being attacked by a parasite or parasitoid.

Host plant — A plant on which an animal feeds, or on which it resides.

Hybrid — Individual resulting from a mating between parents that are genetically unalike. Often used to describe the offspring of matings between the individuals of two different species.

Hybrid breakdown — A reproductive barrier between two species due to the decreased fitness of hybrids compared to either parental species, resulting from the mixing of different adaptations.

Hybrid infertility — A reproductive barrier between two species due to the sterility of any hybrids produced. Although hybrids may develop and reach reproductive maturity, they fail to produce functional gametes.

Hybrid inviability — A reproductive barrier between two species due to the death of hybrids.

Imago — The adult stage in the insect life.

Inbreeding — Reproduction between close relatives.

Inbreeding depression — The reduction in fitness of offspring from matings between close relatives, resulting from the increased likelihood of the same deleterious recessive genes being inherited from both parents and thus being expressed in their progeny.

Indirect selection — In sexual selection, when the mating choice made by an individual (usually the female) leads to an increase in the fitness of its progeny.

Instar — The stage in an insect's life history between two molts. An insect that has recently hatched from an egg and that has not yet molted is said to be a first instar nymph or larva. The adult (imago) is the final instar.

Intersex — An individual of a dioecious species whose primary sex organs and/or secondary sexual traits are partly one sex and partly the other sex, without showing genetically different parts.

Intersexual selection — Selection resulting from variation in the competitive ability of individuals of one sex to attract and be mated by members of the opposite sex.

Intracellular — Within a cell.

Intragenomic conflict — The situation whereby the action of one gene, by increasing its chances of transmission, acts in conflict with the interests of other parts of the genome, leading to selection for suppression of its action.

Intrasexual selection — Selection resulting from variation in the competitive ability of members of one sex (usually males) to gain access to the other sex.

Inversion — State of a chromosome, in which a segment of that chromosome has been reversed.

Invertebrate — An animal lacking a backbone.

Isogamy — Situation in which gametes that fuse at fertilization are morphologically identical.

Jumping gene — *See* transposable element.

Karyotype — The chromosome complement, in terms of number, size, and constitution, of a cell or an organism.

Kin selection — Selection acting on an individual in favor of survival and reproduction not of that individual, but of its relatives, which carry many of the same genes.

Larva — The second stage in the life cycle of insects that undergo full metamorphosis.

Lek — A site where males (or occasionally females) of a species congregate in dense groups and are visited by females (or occasionally males) for the sole purpose of mating.

Lethal gene — A gene that, if expressed, is fatal to the individual bearing it before the individual reaches reproductive maturity.

Locus — The position of a gene on a chromosome.

Major histocompatibility complex — A large cluster of genes that encodes the major histocompatibility antigens in mammals.

Male competition — One of Darwin's two mechanisms of sexual selection. Males compete among themselves (e.g., by aggressive displays or fighting) to gain access to females.

Maternal inheritance — Of traits inherited from the female parent.

Mating plug — Structure placed by a male into the genital opening of a female during copulation to prevent the female from remating.

Mating type — Generally referring to isogamous organisms, this term denotes which individuals can mate. Individuals of the same mating type do not usually mate.

Matriline — Male or female descendents of a single female, connected to her through her female descendents.

Meiosis — The process of two nuclear divisions and a single replication of the chromosome complement of a cell by which the number of chromosomes in resultant daughter cells is reduced by one-half during gamete production.

Meiotic drive — Any type of abnormal meiosis that results in a deviation from expected Mendelian segregation ratios in heterozygotes.

Meiotic spindle — A collection of microtubular fibers to which chromatids attach (at the centromere) during meiosis and that are involved in the segregation of chromatids into daughter cells.

Metamorphosis — An abrupt change in the form of an organism during development that fundamentally alters the function of that organism.

Metaphase plate — The equatorial region of a dividing cell where the chromosomes line up during metaphase.

Mimicry — The resemblance shown by one organism to another for protective or, more rarely, aggressive purposes.

Mitochondria — Organelles in the cytoplasm of plant and animal cells where oxidative phosporylation takes place to produce ATP.

Mitochondrial DNA — The circular, double-stranded genome of eukaryotic mitochondria.

Mitosis — The normal process of cell division in growth, involving the replication of chromosomes, and the division of the nucleus into two, each with the identical complement of chromosomes to the original cell.

Modifier genes — A series of genes, each with small effects, which influences the exact expression of a major gene.

Molecular clock — A theoretical clock based on the premise that the rates at which nucleotide (or amino acid) substitutions become fixed in evolutionary lineages is approximately constant for a given DNA sequence (or polypeptide chain) and reflects the time since taxa diverged.

Monandry — Situation in which females mate with a single male partner over a set period of time.

Monoecious — Of plants in which the reproductive organs of the male and female sex occur on the same individual.

Monogamy — Situation in which an individual has a single mating partner over a set time period.

Monophyletic — Of a taxon that comprises a single clade of organisms.

Morphology — The form and structure of an organism.

Mosaic — The presence in an individual of areas of two genetically different patterns.

Muller's ratchet — The theoretical proposition that populations lacking recombination will accumulate slightly deleterious mutations.

Multiple alleles — A series of forms of a gene occurring at the same locus on a chromosome that have arisen by mutation.

Mutation — Any change in the genetic material, particularly that controlling a particular character or characters, of an organism. Such a change may be due to a change in the number of chromosomes, an alteration in the structure of a chromosome, or a chemical or physical change in an individual gene.

Mutualism — Relationship between two species that benefits both.

Natural selection — According to Charles Darwin, the main mechanism giving rise to evolution. The mechanism by which heritable traits that increase an organism's chances of survival and reproduction are more likely to be passed on to the next generation than less advantageous traits.

Neo-Darwinian theory — A development of Darwin's evolutionary theory refined by incorporating modern biological knowledge, particularly Mendelian genetics, during the mid-20th century.

Neutral mutation — A mutation that is selectively neutral, i.e., it has no adaptive significance.

Nuclear genome — In eukaryotes, the genetic information encoded by all DNA in the nucleus (autosomes, sex chromosomes, and B chromosomes) as opposed to those in other organelles or cytoplasmic elements.

Nucleotide base — The structural unit of a nucleic acid. The major nucleotide bases in DNA are the purines adenine and guanine and the pyrimidines cytosine and thymine, the last of which is replaced by uracil in RNA.

Nucleus — That portion of a cell that contains the chromosomes.

Oocyte — A cell that, on undergoing meiosis, gives rise to an egg.

Organelle — Any substructure within a eukaryotic cell that has a specialized function (e.g., nucleus, mitochondria, chloroplasts, etc.).

Outbreeding — Mating with unrelated individuals in the population.

Oviposit — Egg laying.

Ovipositor — The egg-laying apparatus of a female insect.

Parasite — An organism that lives in or on another organism, its host, obtaining resources at the host's expense.

Parasitoid — An organism (usually an insect) that lays its eggs inside another insect species, where the parasitoid develops, ultimately killing its host.

Parthenogenesis — Reproduction in which eggs develop without fertilization by a male gamete.

Paternal inheritance — Of traits inherited from the male parent.

Pathogen — Any disease-causing organism, but usually referring to microorganisms.

PCR — *See* **Polymerase chain reaction**.

Phenotype — The observable properties of an organism, resulting from the interaction between the organism's genotype and the environment in which it develops.

Pheromone — Chemical substance, which, when released or secreted by an animal, influences the behavior or development of other individuals of the same species.

Photosynthesis — Utilization by plants of light energy to combine carbon dioxide and water into simple sugars.

Phylogenetic — Relating to the pattern of evolutionary descent.

Phylogenetic tree — A diagrammatic representation of genetic distances between populations, species, or higher taxa, the branching of which is said to resemble a tree.

Phylogeny — An evolutionary tree showing the inferred relationships of descent and common ancestry of any given taxon.

Polyandry — Situation in which females may have more than one mating partner, while males have a single mate during a breeding season.

Polygamy — When organisms of either or both sexes have more than one mating partner within a single breeding season.

Polygenic trait — A characteristic controlled polygenically is affected by a large number of genes, each of which has a small effect.

Polygyny — Situation in which males may have more than one mating partner, while females have a single mate during a breeding season.

Polymerase chain reaction (PCR) — A method of amplifying specific DNA sequences by means of repeated rounds of primer-directed DNA synthesis.

Polymorphism — The occurrence of two or more distinctly different forms of a species in the same population.

Polypeptide — A linear molecule with two or more amino acids and one or more peptide groups.

Polyploid — Of individuals or cells, having more than two complete chromosome sets (e.g., triploid = three sets; tetraploid = four sets; hexaploid = six sets, etc.).

Population — A group of organisms of the same species living and breeding together.

Postzygotic reproductive isolation mechanism — Any characteristic of an organism that prevents gene flow between it and members of other species by causing sterility, decreased viability, and/or decreased fitness in zygotes formed as a result of fertilization between gametes of the organism and those of another species.

Prezygotic reproductive isolation mechanism — Any characteristic of an organism that prevents gene flow between it and members of other species by reducing the likelihood of zygote formation, usually by decreasing the likelihood of copulation between the organism and a member of another species.

Primary host — A host that harbors the sexual stage of the life cycle of a parasite or disease organism.

Primary sex organs — The gonads (testes and ovaries), their ducts, and their associated glands.

Primary sex ratio — The sex ratio at fertilization or zygote formation.

Prokaryotes — Single-celled organisms that lack a membrane-bound nucleus or any membrane-bound organelles.

Promiscuity — Mating with many individuals within a population, generally without the formation of strong pair bonds.

Pronotum — The dorsal surface of the first thoracic segment.

Protandry — Course of development of an individual during which its sex changes from male to female.

Proteins — Complex nitrogenous compounds whose molecules consist of numerous amino-acid molecules linked together.

Protogyny — Course of development of an individual during which its sex changes from female to male.

Pseudoautosomal region — A gene located on the homologous region shared by the X and Y sex chromosomes. This region pairs up at meiosis so that the genes in this region behave in the same way as autosomal genes.

Pupa — The third stage in the life history of insects undergoing complete metamorphosis, during which the larval body is rebuilt into that of the adult.

Puparium — The barrel-shaped case of the pupae of some groups of insects, particularly true flies (Diptera).

Random genetic drift — Fluctuation in the frequencies of neutral genes and neutral alleles in a population due to the fact that each generation is only a sample of the one it replaces (same as genetic drift).

Recessive — A recessive allele is not expressed phenotypically when present in a heterozygote, but only when in a homozygote. The opposite of dominant.

Recombination — The production of new combinations of alleles during meiosis (gamete formation), both crossing over of chromosomes and the independent segregation of chromosomes being involved.

Red Queen hypothesis — In Lewis Carroll's *Through the Looking Glass*, the Red Queen states that one has to keep running in order to remain in the same place. Evolutionary arms races are regarded in the same way, in that there is no obvious endpoint to the competition, yet there is constant change.

Reduction division — Phase of meiosis in which homologous chromosomes separate.

Reinforcement — The evolution of reproductive isolation mechanisms between two partially reproductively isolated populations through selection to lower the frequency of hybrid matings.

Replication — The duplication of genomic DNA, or RNA, as part of the processes of mitosis and meiosis.

Reproductive isolation — The situation in which individuals in a group of organisms breed with one another, but not with individuals of other groups. In sexually reproducing organisms, true species are reproductively isolated from one another.

Ribonucleic acid (RNA) — A polynucleotide consisting of a chain of sugar and phosphate units to which are attached various nitrogenous bases (adenine, cytosine, guanine, and uracil). This macromolecule has many diverse functions.

Ribosomes — Multicomponent structures, or organelles, found in all cells, that have a role in transcription, acting as a site of protein synthesis.

RNA — *See* **Ribonucleic acid**.

Sampling error — The random statistical deviations that occur when drawing measurement information from a sample that is supposedly representative of a larger set.

Secondary host — A host that harbors an asexual stage of the life cycle of a parasite or disease organism.

Secondary sex ratio — Of changes in the sex ratio of a family or population that occur after fertilization but prior to reproductive maturity of the individuals involved.

Secondary sexual characteristics — Characteristics that differ between the two sexes, excluding the primary sex organs.

Segregation — The separation from one another of the pairs of genes constituting allelomorphs and their passage, respectively, into different reproductive cells.

Segregation distorter — Any genetic mechanism that leads to a systematic bias in the representation of one or other allele of a heterozygote in the functional gametes produced by an individual.

Sensory exploitation hypothesis — The hypothesis that mate choice (particularly female choice) may evolve out of a pre-existing bias in the sensory biology of an organism.

Sex chromosome — A chromosome that is present in a reproductive cell (or gamete) and that carries the factor for producing a male or female offspring. Such chromosomes are usually denoted by the letters X and Y (or Z and W). In humans, females carry two X chromosomes and males carry one X and one Y chromosome.

Sex-limited inheritance — Inherited traits due to genes, situated in any chromosome, which are expressed in one sex only, although transmitted by both.

Sex linkage — The association of characteristics with sex, because the genes controlling them are situated on the sex chromosomes.

Sex role reversal — Of species in which the operational sex ratio is female biased, so that males are the limiting sex. This leads to female competition for males and male choice of females.

Sexual dimorphism — The existence within a species of morphological differences between the sexes.

Sexual mosaic — An individual of a dioecious species that has some male tissues and some female tissues.

Sexual selection — Selection that promotes traits that will increase an organism's success in mating and ensuring that its gametes are successful in fertilization. This is distinct from natural selection, which simply acts on traits that influence fecundity and survival.

"Sexy sons" — According to Fisher's theory of the evolution of female choice, females who choose to mate with males carrying a particular trait will produce "sexy sons": progeny who tend to carry the preferred trait and so be attractive to other females with the same mating preference.

Sibling — Brother or sister.

Sibling species — Species that are morphologically identical or very similar, but that are genetically different and do not hybridize.

Speciation — The evolution of species.

Species — Groups of actually or potentially interbreeding natural populations that are reproductively isolated from other such groups.

Spermatheca — A small, sac-like branch of the female reproductive tract (of insects and other arthropods) in which sperm may be stored.

Spermatophore — A packet of sperm.

Sperm competition — The competition between sperm from two or more males, within a single female, for fertilization of the ova.

Sperm exhaustion — Situation in which an already mated female has used up the sperm from previous copulations.

Supergene — When a complex set of related traits are controlled by genes that lie so close to each other on the same chromosome that they behave as a single unit.

Superparasitism — The occurrence together of individuals from more than one parent parasitoid within the body of a single host.

Suppressor — Any secondary mutation that completely or partly restores a function lost as a result of the presence of another gene or genetic element.

Symbiont — An organism living in intimate association with another dissimilar organism. Symbiotic associations can be mutualistic, neutral, or parasitic.

Tangled bank — A theory of the evolution of sex based on the advantages of diversity of offspring in an environment with many niches and strong competition for resources.

Taxonomy — The study of the classification of organisms.

Tertiary sex ratio — The sex ratio of a group or population at reproductive maturity.

Tetraploid — An organism that contains four haploid sets of chromosomes.

Thelytoky — The form of parthenogenesis in which unfertilized eggs develop into females.

Tradeoff — Situation in which the allocation of resources to one component of fitness yields a benefit with respect to that component, but incurs a cost with respect to some other component.

Transcription — The process through which RNA is formed along a DNA template. The enzyme RNA polymerase catalyzes the formation of RNA.

Transcription factor — A protein that regulates the transcription of genes.

Translocation — A chromosomal mutation characterized by the change in position of chromosome segments within the chromosome complement.

Transposable element — A piece of DNA that can move from one position in the genome to another.

Transposition — A change in the position of a chromosomal segment without reciprocal exchange.

Triploid — An organism that contains three haploid sets of chromosomes.

Twofold cost of sex — The difference in growth rate between a sexual population where one sex does not provision offspring and a comparable asexual population.

Ultraselfish genes — Genes whose spread and maintenance occur despite and because they cause damage to the individual in which they occur.

Uniparental inheritance — In eukaryotes, the transmission of genetic elements (particularly those of organelles such as mitochondria, ribosomes, and chloroplasts) from only one parent (*see* **Biparental inheritance**).

Unstable equilibrium — The particular state of a system on which forces are precisely balanced, but away from which the system moves when displaced.

Ventral — Concerning the front or underside of an animal.

Vertical transmission — Transmission of genetic elements from parents into progeny.

Viability — The ability of an organism to survive from zygote formation to reproductive maturity.

X chromosome — The chromosome carrying genes concerned with sex determination. There are usually two X chromosomes in one sex and a single X chromosome in the other.

Y chromosome — The partner of the X chromosome in one of the two sexes.

Zygote — The first cell of a new organism, usually resulting from the fusion of two gametes.

R E F E R E N C E S

Ackery, P. R., and Vane-Wright, R. I. 1984. *Milkweed Butterflies*. London: British Museum (Natural History).

Adams, J., Greenwood, P., and Naylor, C. 1987. Evolutionary aspects of environmental sex determination. *International Journal of Invertebrate Reproductive Development*, **11**, 123–36.

Amundsen, T. 2000. Why are female birds ornamental? *Trends in Ecology and Evolution*, **15**, 149–55.

Andrade, M.C.B. 1996. Sexual selection for male sacrifice in the Australian redback spider. *Science*, **271**, 70–72.

Arakaki, N., Miyoshi, T., and Noda, H. 2000. Parthenogenesis-inducing *Wolbachia* in a predatory thrip, *Franklinothrips vespiformis*. *First International Wolbachia Conference Program and Abstracts*, 100–1.

Arnqvist, G., and Henriksson, S. 1997. Sexual cannibalism in the fishing spider and a model for the evolution of sexual cannibalism based on genetic constraints. *Evolutionary Ecology*, **11**, 253–71.

Banks, C. J. 1955. An ecological study of Coccinellidae associated with *Aphis fabae* Scop. on *Vicia faba*. *Bulletin of Entomological Research*, **46**, 561–87.

———. 1956. Observations on the behaviour and early mortality of coccinellid larvae before dispersal from egg shells. *Proceedings of the Royal Entomological Society of London, Series A*, **31**, 56–61.

Barton, N. H., and Charlesworth, B. 1998. Why sex and recombination? *Science*, **281**, 1986–90.

Battaglia, B. 1963. Deviation from panmixia as a consequence of sex determination in the marine copepod *Tisbe reticulata*. *Genetics Today*, **1**, 926.

Bechnel, J. J., and Sweeney, A. W. 1990. *Amblyospora trinus* n. sp. (Microsporida: Amblyosporidae) in the Australia mosquito *Culex halifaxi* (Diptera: Culcidae). *Journal of Protozoology*, **37**, 584–92.

Bell, G. 1982. *The Masterpiece of Nature: The Evolution and Genetics of Sexuality*. San Francisco: University of California Press.

Bernstein, H., Byerly, H. C., Hopf, F. A., and Michod, R. E. 1985. Genetic damage, mutation, and the evolution of sex. *Science*, **229**, 1277–81.

Bernstein, H., Hopf, F.A., and Michod, R.E. 1988. Is meiotic recombination an adaptation for repairing DNA, producing genetic variation, or both? In *The Evolution of Sex* (R. E. Michod and B. R. Levin, eds.), pp. 139–60. Sunderland, MA: Sinauer.

Beukeboom, L. W., and Werren, J. H. 1992. Population genetic analysis of a parasitic chromosome: Experimental analysis of psr in subdivided populations. *Evolution*, **46**, 1257–68.

Blackman, R. 1974. *Aphids*. London: Ginn and Co. Ltd.

Bouchon, D., Rigaud, T., and Juchault., P. 1998. Evidence for widespread *Wolbachia* infection in isopod crustaceans: Molecular identification and host feminization. *Proceedings of the Royal Society of London, Series B*, **265**, 1081–90.

Bouchon, D., Cordaux, R., and Michel-Salzat, A. 2000. *Wolbachia* diversity in crustaceans. *First International Wolbachia Conference Program and Abstracts*, 26–27.

Bourke, A.F.G., and Franks, N. R. 1995. *Social Evolution in Ants*. Princeton, NJ: Princeton University Press.

Bourtzis, K., and O'Neill, S. L. 1998. *Wolbachia* infections and arthropod reproduction. *Bioscience*, **48**, 287–93.

Bourtzis, K., Savakis, C., Ouzounis, C., Kurland, C., Andersson, S., Garrett, R., Braig, H., Ashburner, M., Stouthamer, R., Martin, G., and Hatzigeorgiou, A-G. 2000. The European *Wolbachia* project: Towards novel biotechnological approaches for control of arthropod pests and modification of beneficial arthropod species by endosymbiotic bacteria. *First International Wolbachia Conference Program and Abstracts*, 75.

Bowen, W. R., and Stern, V. M. 1966. Effect of temperature on the production of males and sexual mosaics in uniparental race of *Trichogramma semifumatum*. *Annals of the Entomological Society of America*, **59**, 823–34.

Breeuwer, J.A.J., Stouthamer, R., Barns, S. M., Pelletier, D. A., Weisburg, W. G., and Werren J. H. 1992. Phylogeny of cytoplasmic incompatibility microorganisms in the parasitoid wasp genus *Nasonia* (Hymenoptera: Pteromalidae) based on 16S ribosomal DNA sequences. *Insect Molecular Biology*, **1**, 25–36.

Bull, J. J. 1983. *Evolution of Sex Determining Mechanisms*. Menlo Park, CA: Benjamin Cummings.

Bulnheim, H. P. 1978. Interactions between genetic, external and parasitic factors in sex determination of the crustacea amphipod *Gammarus duebeni*. *Hegolander Wissenschaftliche Meeresuntersuchungen*, **31**, 1–33.

Burt, A. 2000. Sex, recombination, and the efficacy of selection—Was Weismann right? *Evolution*, **54**, 337–51.

Cabello, T., and Vargas, P. 1985. Temperature as a factor influencing the form of reproduction of *Trichogramma cordubensis*. *Zeitschrift für Angewandte Entomologie*, **100**, 434–41.

Carroll, L. 1962. *Through the Looking-glass and What Alice Found There*. London: Macmillan and Co Ltd.

Chanter, D. O., and Owen, D. F. 1972. The inheritance and population genetics of sex ratio in the butterfly *Acraea encedon*. *Journal of Zoology*, **166**, 363–83.

Charlesworth, B. 1978. Model for evolution of Y chromosomes and dosage compensation. *Proceedings of the National Academy Sciences*, **75**, 5618–22.

———. 1991. The evolution of sex chromosomes. *Science*, **251**, 1030–33.

Charnov, E. L., Los-den Hartogh, R. L., Jones, W. T., and van den Assem, J. 1981. Sex ratio evolution in a variable environment. *Nature*, **289**, 27–33.

Chesley, P., and Dunn, L. C. 1936. The inheritance of taillessness (anury) in the house mouse. *Genetics*, **21**, 525–36.

Clarke, C. A., Sheppard, P. M. and Scali, V. 1975. All female broods in the butterfly *Hypolimnas bolina*. *Proceedings of the Royal Society of London, Series B*, **189**, 29–37.

Cline, T. W. 1993. The *Drosophila* sex determination signal: How do flies count to two? *Trends in Genetics,* **9**, 385–90.

Clutton-Brock, T. H., Albon, S. D., and Guinness, F. E. 1984. Maternal dominance, breeding success, and birth sex ratio in red deer. *Nature*, **308**, 358–60.

Cobb, J. A. 1914. Human fertility. *Eugenics Review*, **4**, 379–82.

Conan Doyle, Sir A. 1979. *A Case of Identity*. In *The Adventures of Sherlock Holmes*. London: Pan Books.

———. 1982. *A Study in Scarlet*. Harmondsworth: Penguin.

Cook, J. M., and Crozier, R. H. 1995. Sex determination and population biology in the Hymenoptera. *Trends in Ecology and Evolution*, **10**, 281–86.

Correns, C. 1909. *Vererbunsversuche mit blass (gelb) grunen und buntblättrigen sippen bei* Mirabilis jalapa, Urtica pilulifera *und* Lunaria annua. *Zeitschrift für Induktive Abstammungs und Vererbungslehre*, **1**, 291–329.

Cosmides, L., and Tooby, J. 1981. Cytoplasmic inheritance and intragenomic conflict. *Journal of Theoretical Biology*, **89**, 83–129.

Counce, S. J., and Poulson, D. F. 1962. Developmental effects of the sex-ratio agent in embryos of *Drosophila willistoni*. *Journal of Experimental Zoology*, **151**, 17–31.

Crow, J. F. 1988. The ultraselfish gene. *Genetics*, **118**, 389–91.

Crozier, R., and Pamilo, P. 1996. *Evolution of Social Insect Colonies: Sex Allocation and Kin Selection*. Oxford: Oxford University Press.

Cunningham, E.J.A., and Russell, A. F. 2000. Egg investment is influenced by male attractiveness in the mallard. *Nature*, **404**, 74–77.

Darwin, C. R. 1859. *On the Origin of Species by Means of Natural Selection, or the Preservation of Favoured Races in the Struggle for Life*. London: John Murray.

———. 1871. *The descent of man and selection in relation to sex*. London: John Murray.

———. 1874. *The descent of man and selection in relation to sex*. 2nd ed. London: John Murray.

———. [1887]. 1958. *The Autobiography of Charles Darwin 1809–1882* (N. Barlow, ed.). London: Collins.

Dawkins, R. 1982. *The Extended Phenotype*. Oxford: Oxford University Press.

Diesel, R. 1990. Sperm competition and reproductive success in the decapod *Inachus phalangium* (Majidae): A male ghost spider crab that seals off rivals' sperm. *Journal of Zoology*, **200**, 213–23.

Dixon, A.F.G. 1998. *Aphid Ecology*. 2nd ed. London: Chapman and Hall.

Doncaster, L. 1913. On the inherited tendency to produce purely female families in *Abraxas grossulariata*, and its relation to an abnormal chromosome number. *Journal of Genetics*, **3**, 1–10.

———. 1914. On the relations between chromosomes, sex limited transmission and sex determination in *Abraxas grossulariata*. *Journal of Genetics*, **4**, 1–21.

Drapeau, M. D. 1999. Local mating and sex ratios. *Trends in Ecology and Evolution*, **14**, 235.

Dunn, A. M., and Rigaud, T. 1998. Horizontal transfer of parasitic sex ratio distorters between crustacean hosts. *Parasitology*, **117**, 15–19.

Düsing, C. 1884. *Die Regulierung des Geschlechtsverhältnisses*. Jena, Germany: Fischer.

Dyson, E. 2000. Inherited parasites in the butterfly *Hypolimnas bolina* (Lepidoptera: Nymphalidae). *First International Wolbachia Conference Program and Abstracts*, 82–83.

Eberhard, W. G. 1996. *Female Control: Sexual Selection by Cryptic Female Choice*. Princeton, NJ: Princeton University Press.

Edwards, A.W.F. 1998. Natural selection and the sex ratio: Fisher's sources. *American Naturalist*, **151**, 564–69.

Ehrlich, A. H., and Ehrlich, P. R. 1978. Reproductive strategies in the butterflies. I. Mating frequency, plugging and egg numbers. *Journal of the Kansas Entomological Society*, **51**, 666–97.

Ehrman, L., and Kernaghan, R. P. 1971. Microorganismal basis of the infectious hybrid sterility in *Drosophila paulistorum*. *Journal of Heredity*, **62**, 66–71.

Elgar, M. A., and Nash, D. R. 1988. Sexual cannibalism in the garden spider, *Aranus diadematus*. *Animal Behaviour*, **36**, 1511–17.

Felsenstein, J. 1974. The evolutionary advantage of recombination. *Genetics*, **78**, 737–56.

———. 1985. Recombination and sex: Is Maynard Smith necessary? In *Evolution: Essays in Honour of John Maynard Smith* (H. Greenwood and M. Slatkin, eds.), pp. 209–220. Cambridge: Cambridge University Press.

Fischer, E. A. 1980. The relationship between mating systems and simultaneous hermaphroditism in the coral reef fish, *Hypoplectus nigricans. Animal Behaviour*, **28**, 620–33.

Fisher, R. A. 1930. *The Genetical Theory of Natural Selection.* Oxford: Oxford University Press.

———. 1931. The evolution of dominance. *Biological Reviews*, **6**, 345–368.

Flanders, S. A. 1945. The bisexuality of uniparental hymenoptera, a function of the environment. *American Naturalist*, **79**, 122–141.

Ford, E. B. 1940. Polymorphism and taxonomy. In *The New Systematics* (J. S. Huxley, ed.), pp. 493–513. Oxford: Clarendon Press.

Fredga, K., Gropp, A., Winking, H., and Frank, F. 1977. A hypothesis explaining the exceptional sex ratio in the wood lemming (*Myopus schisticolor*). *Hereditas*, **85**, 101–4.

Fricke, H. W., and Fricke, S. 1977. Monogamy and sex change by aggressive dominance in coral reef fish. *Nature*, **266**, 830–32.

Geier, P. W., and Briese, D. T. 1980. The light brown apple moth, *Epiphyas postvittana* (Walker). 4. Studies on population dynamics and injuriousness to apples in Australian Capital territory. *Australian Journal of Ecology*, **5**, 63–93.

Geoghegan, I. E., Majerus, T.M.O., and Majerus, M.E.N. 1998. A record of a rare male of the parthenogenetic parasitoid *Dinocampus coccinellae* (Schrank) (Hym., Braconidae). *Entomologist's Record and Journal of Variations*, **110**, 171–72.

Ghelelovitch, S. 1952. *Sur le déterminisme génétique de la stérilité dans les croisements entre différentes souches de* Culex autogenicus *Roubaud. Comptes Rendus de l'Academie des Sciences, Paris, Série III*, **234**, 2386–88.

Gilbert, L. E. 1976. Postmating female odour in *Heliconius* butterflies: A male contributed aphrodisiac. *Science*, **193**, 419–20.

Gileva, E. A. 1987 Meiotic drive in the sex chromosome system of the varying lemming *Dicrostonyx torquatus* Pall. (Rodentia, Microtinae). *Heredity*, **59**, 383–389.

Ginsburger-Vogel, T., and Desportes, I. 1979. Structure and biology of *Marteilia* sp. in the amphipod *Orchestia gammarellus. Marine Fisheries Review*, **41**, 3–7.

Godfray, H.C.J., and Werren, J. H. 1996. Recent developments in sex ratio studies. *Trends in Ecology and Evolution*, **11**, 59–63.

Goldschmidt, R. B. 1934. *Lymantria. Bibliographia Genetica*, **11**, 1–186.

———. 1940. *The Material Basis of Evolution.* New Haven, CT: Yale University Press.

Gordon, I. J. 1984. Mimicry, migration and speciation in *Acraea encedon* and *Acraea encedana.* In *The Biology of Butterflies* (R. I. Vane-Wright and P. R. Ackery, eds.), pp. 193–196. London: Academic Press.

Gorman, M., and Baker, B. S. 1994. How flies make one equal two: Dosage compensation in *Drosophila. Trends in Genetics*, **10**, 376–80.

Haig, D., and Grafen, A. 1991. Genetic scrambling as a defence against meiotic drive. *Journal of Theoretical Biology*, **153**, 531–58.

Haldane, J.B.S. 1932. *The Causes of Evolution.* New York: Harper.

———. 1955. Population genetics. *New Biology*, **18**, 34–51.

Hamilton, W. D. 1964. The genetical evolution of social behaviour. *Journal of Theoretical Biology*, **7**, 1–52.

———. 1967. Extraordinary sex ratios. *Science*, **156**, 477–88.

Hamilton, W. D., and Zuk, M. 1982. Heritable true fitness and bright birds: A role for parasites? *Science*, **218**, 384–87.

Hamilton, W. D., Axelrod, R., and Tanese, R. 1990. Sexual reproduction as an adaptation to resist parasites (a review). *Proceedings of the National Academy of Sciences USA*, **87**, 3566–73.

Harshman, L. G., and Prout, T. 1994. Sperm displacement without sperm transfer in *Drosophila melanogaster*. *Evolution*. **48**, 758–66.

Hastings, I. A. 1994. Manifestations of sexual selection may depend on the genetic basis of sex determination. *Proceedings of the Royal Society of London, Series B*, **258**, 83–87.

Hatcher, M. J., Dunn, A. M., and Tofts, C.M.N. 1997. The effect of the embryonic bottleneck on vertically transmitted parasites. In *Computation in Cellular and Molecular Biological Systems* (R. Cuthbertson, N. Holcombe, and R. Paton, eds.), pp. 339–51. Singapore: World Scientific.

Hickey, W. A., and Craig, G.B.J. 1966. Genetic distortion of sex ratio in a mosquito, *Aedes aegypti*. *Journal of Genetics*, **53**, 1177–96.

Higashiura, Y., Ishihara, M., and Schaefers, P. W. 1999. Sex ratio distortion and severe inbreeding depression in the gypsy moth, *Lymantria dispar* L. in Hokkaido, Japan. *Heredity*, **83**, 290–97.

Hoffmann, A. A., and Turelli, M. 1997. Cytoplasmic incompatibility in insects. In *Influential Passengers* (S. L. O'Neill, A. A. Hoffmann, and J. H. Werren, eds.), pp. 42–80. Oxford: Oxford University Press.

Hoffmann, A. A., Turelli, M., and Harshman, L. G. 1990. Factors affecting the distribution of cytoplasmic incompatibility in *Drosophila simulans*. *Genetics*, **126**, 933–48.

Holland, B., and Rice, W. R. 1999. Experimental removal of sexual selection reverses intersexual antagonistic coevolution and removes a reproductive load. *Proceedings of the National Academy of Sciences USA*, **96**, 5083–88.

Hosken, D. J. 1999. Sperm replacement in yellow dung flies: A role for females. *Trends in Ecology and Evolution*, **14**, 151–52.

Huigens, M.E.H., Luck, R. F., Klaassen, R.G.H., Maas, F.M.P.M., Timmermans, M.J.T.N., and Stouthamer, R. 2000. Infectious parthenogenesis. *Nature*, **405**, 178–79.

Hurst, G.D.D., and Majerus, M.E.N. 1993. Why do maternally inherited micro-organisms kill males? *Heredity* **71**, 81–95.

Hurst, G.D.D., and McVean, G.A.T. 1998. Parasitic male-killing bacteria and the evolution of clutch size. *Ecological Entomology*, **23**, 350–53.

Hurst, G.D.D., Purvis, E. L., Sloggett, J. J., and Majerus, M.E.N. 1994. The effect of infection with male-killing *Rickettsia* on the demography of female *Adalia bipunctata* L. (two-spot ladybird). *Heredity*, **73**, 309–16.

Hurst, G.D.D., Sloggett, J. J., and Majerus, M.E.N. 1996. Estimation of the rate of inbreeding in a natural population of *Adalia bipunctata* (Coleoptera: Coccinellidae) using a phenotypic indicator. *European Journal of Entomology*, **93**, 145–50.

Hurst, G.D.D., Hurst, L. D., and Majerus, M.E.N. 1997. Cytoplasmic sex-ratio distorters. In *Influential Passengers* (S. L. O'Neill, A. A. Hoffmann, and J. H. Werren, eds.), pp. 125–54. Oxford: Oxford University Press.

Hurst, G.D.D., Jiggins, F. M., Schulenberg, J.H.G.v.d., Bertrand, D., West, S. A., Goriacheva, I. I., Zakharov, I. A., Werren, J. H., Stouthamer, R., and Majerus, M.E.N. 1999. Male-killing *Wolbachia* in two species of insect. *Proceedings of the Royal Society of London, Series B*, **266**, 735–40.

Hurst, L. D. 1990. Parasite diversity and the evolution of diploidy, multicellularity and anisogamy. *Journal of Theoretical Biology*, **144**, 429–33.

———. 1991. The incidences and evolution of cytoplasmic male-killers. *Proceedings of the Royal Society of London, Series B*, **244**, 91–99.

———. 1993. The incidences, mechanisms and evolution of cytoplasmic sex ratio distorters in animals. *Biological Reviews*, **68**, 121–93.

Hurst, L. D., and Hamilton, W. D. 1992. Cytoplasmic fusion and the nature of sexes. *Proceedings of the Royal Society of London, Series B*, **247**, 189–94.

Hurst, L. D., and Pomiankowski, A. N. 1991. Causes of sex ratio bias may account for unisexual sterility in hybrids: A new explanation of Haldane's rule and related phenomena. *Genetics*, **128**, 841–58.

Isawa, Y., Pomiankowski, A., and Nee, S. 1991. The evolution of costly mate preferences. II. The 'Handicap principle.' *Evolution*, **45**, 1431–42.

Jackson, D. J. 1958. Observations on the biology of *Caraphractus cinctus* Walker (Hymenoptera: Mymaridae), a parasitoid of the eggs of Dytiscidae. I. Methods of rearing and numbers bred on different host eggs. *Transactions of the Royal Entomological Society of London*, **110**, 533–54.

Jeyaprakash, A., and Hoy, M. A. 2000. Long PCR improves *Wolbachia* DNA amplification: wsp sequences found in 76% of sixty-three arthropod species. *Insect Molecular Biology*, **9**, 393–405.

Jiggins, F. M. 2000. *The Causes and Consequences of Sex Ratio Distortion in African Butterflies*. Ph. D. thesis: University of Cambridge.

Jiggins, F. M., Hurst, G.D.D., and Majerus, M.E.N. 1998. Sex-ratio distortion in *Acraea encedon* (Lepidopera: Nymphalidae) is caused by a male-killing bacterium. *Heredity*, **81**, 87–91.

———. 1999a. How common are meiotically driving sex chromosomes in insects? *American Naturalist*, **154**, 841–43.

———. 1999b. Sex ratio distorting *Wolbachia* causes sex role reversal in its butterfly host. *Proceedings of the Royal Society of London, Series B*, **267**, 69–73.

Jiggins, F. M., Hurst, G.D.D., Schulenburg, J.H.G.v.d., and Majerus, M.E.N. 2000a The butterfly *Danaus chrysippus* is infected with a male-killing *Spiroplasma* bacterium. *Parasitology*, **120**, 439–46.

Jiggins, F. M., Hurst, G.D.D., Dolman, C. E., and Majerus, M.E.N. 2000b. High prevalence male-killing *Wolbachia* in the butterfly *Acraea encedana. Journal of Evolutionary Biology*, **13**, 495–501.

Jiggins, F. M., Schulenburg, J.H.G.v.d., Hurst, G.D.D., and Majerus, M.E.N. 2001a. Recombination confounds interpretations of *Wolbachia* evolution. *Proceedings of the Royal Society of London, Series B*, **268**, 1423–27.

Johnson, M. S., and Turner, J.R.G. 1979. Absence of dosage compensation for a sex-linked enzyme in butterflies (*Heliconius*). *Heredity*, **43**, 71–77.

Johnstone, R. A. and Hurst, G.D.D. 1996. Maternally inherited male-killing microorganisms may confound interpretation of mtDNA variation in insects. *Biological Journal of Linnean Society*, **58**, 453–70.

Johnstone, R. A., Reynolds, J. D., and Deutsch, J. C. 1996. Mutual mate choice and sex differences in choosiness. *Evolution*, **50**, 1382–91.

Juchault, P., and Legrand, J. J. 1972. *Croisements de néo-mâles expérimentaux chez* Armadillium vulgare *Latr. (Crustacé, Isopode, Oniscoïde). Mise en évidence d'une hétérogamétie femelle. Comptes Rendus de l'Academie des Sciences, Paris, Série III*, **274**, 1387–89.

———. 1985. *Contribution à l'étude du mécanisme de l'état réfractaire à l'hormone androgène chez les* Armadillium vulgare *Latreille (Crustacé, Isopode, Oniscoïde) hébergeant une bactérie féminisante. General and Comparative Endocrinology*, **60**, 463–67.

Juchault, P., and Rigaud, T. 1995. Evidence of female heterogamety in two terrestrial crustaceans and the problems of sex chromosome evolution in isopods. *Heredity*, **75**, 466–71.

Juchault, P., Rigaud, T., and Mocquard, J. P. 1993. Evolution of sex determination and sex ratio variability in wild populations of *Armadillium vulgare* Latr. (Crustacea, Isopoda): A case study in conflict resolution. *Acta Oecologica*, **14**, 547–62.

Kageyama, D., Hoshizaki, S., and Ishikawa, Y. 1998. Female-biased sex ratio in the Asian corn borer, *Ostrinia furnacalis*: Evidence for the occurrence of feminizing bacteria in an insect. *Heredity*, **81**, 311–16.

Kageyama, D., Nishimura, G., Hoshizaki, S., and Ishikawa, Y. 2000. *Wolbachia* infection causes feminization in two species of moths, *Ostrinia furnacalis* and *O. scapulalis* (Lepidoptera: Crambidae). *First International Wolbachia Conference Program and Abstracts*, 106–07.

Khalil, A. M., and Murakami, N. 1999. Factors affecting the hareem formation process by young Misaki feral stallions. *Journal of Veterinary Medical Science*, **61**, 667–71.

King, B. H., and Skinner, S. W. 1991. Sex ratio in a new species of *Nasonia* with fully-winged males. *Evolution*, **45**, 225–28.

Klarwill, V. von, ed. 1924. *The Fugger News-Letters: Being a Selection of Unpublished Letters from the Correspondents of the House of Fugger during the Years 1568–1605.* London: John Lane the Bodley Head Ltd.

Kondrashov, A. S. 1982. Selection against harmful mutations in large sexual and asexual populations. *Genetical Research*, **40**, 325–32.

———. 1984. Deleterious mutations as an evolutionary factor. I. The advantage of recombination. *Genetical Research*, **44**, 199–217.

Krasfur, E. S., Whitten, C. J., and Novy, J. E. 1987. Screwworm eradication in North and Central America. *Parasitology Today*, **3**, 131–37.

Krebs, J. R., and Davies, N. B. 1981. *An Introduction to Behavioural Ecology*. Oxford: Blackwell Scientific Publications.

———. 1987. *An Introduction to Behavioural Ecology*. 2nd ed. Oxford: Blackwell Scientific Publications.

Laven, H. 1967. Speciation and evolution in *Culex pipiens*. In *Genetics of Insect Vectors of Disease* (J. Wright and R. Pal, eds.), pp. 251–75. Amsterdam: Elsevier.

Lawson, E. T., Mousseau, T. A., Klaper, R., Hunter, M. D., and Werren, J. H. 2001. Rickettsia associated with male-killing in a buprestid beetle. *Heredity*, **86**, 497–505.

Legrand, J. J., and Juchault, P. 1970. *Modification expérimentale de la proportion des sexes chez les Crustacés Isopodes terrestres: Induction de la thélygénie chez* Armadillium vulgare *(Latr.). Comptes Rendus de l'Academie des Sciences, Paris, Série III*, **270**, 706–8.

Levings, C. S. III, and Pring, D. R. 1976. Restriction endonuclease analysis of mitochondrial DNA from normal and Texas cytoplasmic male sterile maize. *Science*. **193**, 158–60.

Lusis, J. J. 1947. Some aspects of the population increase in *Adalia bipunctata* 2. The strains without males. *Doklady Akademii Nauk SSSR (Moskva)*, **57**, 825–28.

Lyttle, T. W. 1991. Segregation distorters. *Annual Review of Genetics*, **25**, 511–57.

MacLellan, C. R. 1973. Natural enemies of the light brown apple moth, *Epiphyas postvittana*, in the Australian capital territory. *Canadian Entomologist*, **105**, 681–700.

Mader, L. 1926–1937. *Evidenz der Palaarktischen Coccinelliden und ihrer Aberationen in Wort und Bild*. Vienna: Troppau.

Magnin, M., Pasteur, N., and Raymond, M. 1987. Multiple incompatibilities within populations of *Culex pipiens* L. in southern France. *Genetica*, **74**, 125–30.

Majerus, M.E.N. 1981. All female broods of *Philudoria potatoria* Linn. (Lepidoptera: Lasciocampidae). *Proceedings of the British Entomological and Natural History Society*, **14**, 97–92.

———. 1994. *Ladybirds. New Naturalist Series 81*. London: HarperCollins.

———. 1998. *Melanism: Evolution in Action*. Oxford: Oxford University Press.

———. 1999. *Simbiontes hereditarios causantes de efectos deletéreos en los artrópodos/* Deleterious endosymbionts of Arthropods. In *The Evolution and Ecology of Arthropods* (A. Melic, J. J. De Haro, M. Méndez, and I. Ribera, eds.), pp. 777–806. (In Spanish and English.) Zaragosa, Spain: Sociedad Entomologica Aragonera.

Majerus, M.E.N., and Hurst, G.D.D. 1996. Extension to the genome. *The Genetical Society Newsletter*, **32**, 6.

———. 1997. Ladybirds as a model system for the study of male-killing symbionts. *Entomophaga*, **42**, 13–20.

Majerus, M.E.N., and Kearns, P.W.E. 1989. *Ladybirds* (Naturalists' Handbooks 10). Slough, UK: Richmond Publishing.

Majerus, M.E.N., and Majerus, T.M.O. 1997. Cannibalism among ladybirds. *Bulletin of the Amateur Entomologists' Society*, **56**, 235–48.

———. 2000a. Female-biased sex ratio due to male-killing in the Japanese ladybird *Coccinula sinensis*. *Ecological Entomology*, **25**, 1–5.

———. 2000b. Nuclear suppression of a male-killing bacterium in a ladybird beetle. *First International Wolbachia Conference Program and Abstracts*, 39–40.

Majerus, M.E.N., Amos, W., and Hurst, G.D.D. 1996. *Evolution: The Four Billion Year War*. Harlow, UK: Longmans.

Majerus, M.E.N., Schulenburg, J.H.G.v.d., and Zakharov, I. A. 2000. Multiple cause of male-killing in a single sample of the 2 spot ladybird, *Adalia bipunctata* (Coleoptera: Coccinellidae) from Moscow. *Heredity*, **84**, 605–9.

Majerus, T.M.O. 2001. *The Evolutionary Genetics of Male-killing in the Coccinellidae*. Ph.D thesis: University of Cambridge.

Majerus, T.M.O., Majerus, M.E.N., Knowles, B., Wheeler, J., Bertrand, D., Kuznetzov, V.N., Ueno H., and Hurst, G.D.D. 1998. Extreme variation in the prevalence of inherited male-killing microorganisms between three populations of the ladybird *Harmonia axyridis* (Coleoptera: Coccinellidae). *Heredity*, **81**, 683–91.

Majerus, T.M.O., Schulenburg, J.H.G.v.d., Majerus, M.E.N., and Hurst, G.D.D. 1999. Molecular identification of a male-killing agent in the ladybird *Harmonia axyridis* (Pallas) (Coleoptera: Coccinellidae). *Insect Molecular Biology*, **8**, 551–55.

Martin, G., Juchault, P., and Legrand, J. J. 1973. *Mise en évidence d'un micro-organisme intracytoplasmique symbionte de l'Oniscoide* Armadillium vulgare *L., dont la présence accompagne l'intersexualité ou la féminisation totale des males génétiques de la lignée thélyghne. Comptes Rendus de l'Academie des Sciences, Paris, Série III*, **276**, 2313–16.

Matsuka, M., Hashi, H., and Okada, I. 1975. Abnormal sex-ratio found in the lady beetle, *Harmonia axyridis* Pallas (Coleoptera: Coccinellidae). *Applied Entomology and Zoology*, **10**, 84–89.

Maynard Smith, J. 1964. Group selection and kin selection. *Nature*, **201**, 1145–47.

———. 1989. *Evolutionary Genetics*. Oxford: Oxford University Press.

Mayr, E., and Provine, W. B. 1980. *The Evolutionary Synthesis*. Cambridge, MA: Harvard University Press.

Miller, D. R., and Borden, J. H. 1985. Life history and biology of *Ips latidens* (Leconte) (Coleoptera: Scolytidae). *Canadian Entomologist*, **117**, 859–71.

Milton, J. 2000. *Paradise Lost.* (John Leonard, ed., for Penguin Classics Series.) London: Penguin.

Moreau, J., Bertin, A., Caubet, Y., and Rigaud, T. 2000. *Wolbachia* infection and female mating success in the isopod *Armadillium vulgare. First International Wolbachia Conference Program and Abstracts*, 108.

Morgan, T. H., Bridges, C. B., and Sturtevant, A. H. 1925. The genetics of *Drosophila. Bibliographia Genetica*, **2**, 1–262.

Morimoto, S., Nakai, M., Ono, A., and Kunimi, Y. 2001. Late male-killing phenomenon found in a Japanese population of the oriental tea tortrix, *Homona magnanima* (Lepidoptera: Tortricidae). *Heredity*, **87**, 435–40.

Muller, H. J. 1932. Some genetic aspects of sex. *American Naturalist*, **66**, 118–38.

Nur, U., Werren, J. H., Eickbush, D. G., Burke, W. D., and Eickbush, T. H. 1988. A selfish B-chromosome that enhances its transmission by eliminating the paternal genome. *Science*, **240**, 512–14.

O'Donald, P. 1980. *Genetic Models of Sexual Selection.* Cambridge, UK: Cambridge University Press.

O'Donald, P., Derrick, M., Majerus, M.E.N., and Wier, J. 1984. Population genetic theory of the assortative mating, sexual selection and natural selection of the two-spot ladybird, *Adalia bipunctata. Heredity*, **52**, 43–61.

O'Neill, S. L., Hoffmann, A. A., and Werren, J. H., eds. 1997. *Influential Passengers.* Oxford: Oxford University Press.

O'Neill, S. L., Eisen, J., Slatko, B., Sun, L., Foster, J., Blaxter, M., and Scott, A. 2000. Comparative genome sequencing of *Wolbachia. First International Wolbachia Conference Program and Abstracts*, 74.

Osawa, N., and Nishida, T. 1992. Seasonal variation in elytral colour polymorphism in *Harmonia axyridis* (the ladybird beetle): The role of non-random mating. *Heredity*, **69**, 297–307.

Owen, D. F. 1965. Change in sex ratio in an African butterfly. *Nature*, **206**, 744.

———. 1970. Inheritance of sex ratio in the butterfly *Acraea encedon. Nature*, **225**, 662–63.

Owen, D. F., and Chanter, D. O. 1968. Population biology of tropical African butterflies. 2. Sex ratio and polymorphism in *Danaus chrysippus* L. *Revue de Zoologie et de Botanique Africaines, Bruxelles*, **78**, 81–97.

———. 1969. Population biology of tropical African butterflies. Sex ratio and genetic variation in *Acraea encedon. Journal of Zoology, London*, **157**, 345–74

Owen, D. F., Smith, D.A.S., Gordon, I. J., and Owiny, A. M. 1994. Polymorphic Müllerian mimicry in a group of African butterflies: A reassessment of the relationship between *Danaus chrysippus, Acraea encedon* and *Acraea encedana* (Lepidoptera: Nymphalidae). *Journal of Zoology, London*, **232**, 93–108.

Pardo, M. C., López-León, M. D., Hewitt, G. M., and Camacho, J.P.M. 1995. Female fitness is increased by frequent mating in grasshoppers. *Heredity*, **74**, 654–60.

Parker, G. A. 1970. Sperm competition and its evolutionary consequences in insects. *Biological Review*, **45**, 525–67.

Parker, G. A., Baker, R. R., and Smith, V.G.F. 1972. The origin and evolution of gamete dimorphism and the male-female phenomenon. *Journal of Theoretical Biology*, **36**, 529–53.

Parkhurst, S. M., and Meneely, P. M. 1994. Sex determination and dosage compensation: Lessons from flies and worms. *Science*, **264**, 924–32.

Perkins, R.C.L. 1905. Leaf-hoppers and their natural enemies (Mymaridae, Platygasteridae). *Hawaii Sugar Planters Association Experimental Station Bulletin*. **1**, 187–203.

Perrot-Minnot, M-J., Guo, L. R., and Werren, J. H. 1996. Single and double infections with *Wolbachia* in the parasitic wasp *Nasonia vitripennis*: Effects on compatibility. *Genetics*, **143**, 961–72.

Pijls, W.A.M., van Steenbergen, H. J., and van Alphen, J.J.M. 1996. Asexuality cured: The relations and differences between sexual and asexual *Apoanagyrus diversicornis*. *Heredity*, **76**, 506–13.

Pitnick, S., Spicer, G. S., and Marhow, T. A. 1995. How long is a giant sperm? *Nature*, **375**, 109.

Pizzari, T., and Birkhead, T. R. 2000. Female feral fowl eject sperm of sub-dominant males. *Nature*, **405**, 787–89.

Policansky, D. 1981. Sex choice and size advantage model in jack-in-the-pulpit (*Arisaema triphyllum*). *Proceedings of the National Academy of Sciences*, **78**, 1306–8.

Poulton, E. B. 1914. W. A. Lamborn's breeding experiments upon *Acraea encedon* (Linn.) in the Lagos district of West Africa, 1910–12. *Journal of the Linnean Society (Zoology)*, **32**, 391–416.

Pringle, J. A. 1938. A contribution to the knowledge of *Micromalthus debilis* Le C. (Coleoptera). *Transactions of the Royal Entomological Society of London*, **87**, 271–86.

Ralph, C. P. 1977. Effect of the host plant density on populations of a specialized seed sucking bug, *Oncopeltus fasciatus*. *Ecology*, **58**, 799–809.

Randerson, J. P., Smith, N.G.C., and Hurst, L. D. 2000. The evolutionary dynamics of male-killers and their hosts. *Heredity*, **84**, 152–60.

Rice, W. R. 1987. The accumulation of sexually antagonistic genes as a selective agent promoting the evolution of reduced recombination between primitive sex chromosomes. *Evolution*, **41**, 911–14.

———. 1992. Sexually antagonistic genes: Experimental evidence. *Science*. **256**, 1436–39.

———. 1994. Degeneration of a nonrecombining chromosome. *Science*, **263**, 230–32.

Richardson, P. M., Holmes, W. P., and Saul, G. B. II. 1987. The effect of tetracycline on nonreciprocal cross incompatibility in *Mormoniella* [= *Nasonia*] *vitripennis*. *Journal of Invertebrate Pathology*, **50**, 176–83.

Ricklefs, R. E. 1990. *Ecology*. 3rd ed. New York: W. H. Freeman.

Ridley, M. 1993. *The Red Queen: Sex and the Evolution of Human Nature*. London: Viking.

Rigaud, T. 1997. Inherited microorganisms and sex determination of arthropod hosts. In *Influential Passengers* (S. L. O'Neill, A. A. Hoffmann, and J. H. Werren, eds.), pp. 81–101. Oxford: Oxford University Press.

Rigaud, T., and Juchault, P. 1992. Genetic control of the vertical transmission of a cytoplasmic sex factor in *Armadillium vulgare* Latr. (Crustacea, Oniscidea). *Heredity*, **68**, 47–52.

Rigaud, T., Souty-Grosset, C., Raimond, R., Mocquard, J. P., and Juchault, P. 1991. Feminizing endocytobiosis in the terrestrial crustacean *Armadillium vulgare* Latr. (Isopoda): Recent acquisitions. *Endocytobiosis and Cell Research*, **7**, 259–73.

Roldan, E.R.S., and Gomendio, M. 1999. The Y chromosome as a battle ground for sexual selection. *Trends in Ecology and Evolution*, **14**, 58–62.

Rössler, Y., and DeBach, P. 1972. The biosystematic relations between a thelytokous and arrhenotokous form of *Aphytis mytilaspidis*. I. The reproductive relations. *Entomophaga*, **17**, 391–423.

Rousset, F., Bouchon, D., Pintureau, B., Juchaut, P., and Solignac, M. 1992. *Wolbachia* endosymbionts responsible for various alterations of sexuality in arthropods. *Proceedings of the Royal Society of London, Series B*, **250**, 91–98.

Ryan, S. L., Saul, G. B., and Conner, G. W. 1985. Aberrant segregation of R-locus genes in male progeny from incompatible crosses in *Mormoniella. Journal of Heredity*, **76**, 21–26.

Sakaguchi, B., and Poulson, D. F. 1963. Interspecific transfer of the "sex ratio" condition from *Drosophila willistoni* to *D. melanogaster. Genetics*, **48**, 841–61.

Sandler, L., and Novitski, E. 1958. Meiotic drive as an evolutionary force. *American Naturalist*, **91**, 105–10.

Sasaki, K., Satoh, T., and Obara, Y. 1996. The honeybee queen has the potential ability to regulate the primary sex ratio. *Applied Entomology Zoology*, **31**, 247–54.

Sasaki, T., and Iwahashi, O 1995. Sexual cannibalism in an orb-weaving spider *Argiope aemula. Animal Behaviour*, **49**, 1119–21.

Sasaki, T., Fujii, Y., Kageyama, D., Hoshizaki, S., and Ishikawa, H. 2000. Interspecific transfer of *Wolbachia* in Lepidoptera: The feminizer of the Adzuki bean borer *Ostrinia scapulalis* causes male-killing in the Mediterranean flour moth *Ephestia kuehniella. First International Wolbachia Conference Program and Abstracts*, 109–10.

Sassamanm, C., and Weeks, S. C. 1993. The genetic mechanism of sex determination in the concoshtrocan shrimp *Eulimnadia texana. American Naturalist*, **141**, 314–28.

Saul, G. B. 1961. An analysis of non reciprocal cross incompatibility in *Mormoniella vitripennis* (Walker). *Zeitschrift für Vererbungslehre*, **92**, 28–33.

Scali, V., and Masetti, I. 1973. The population structure of *Maniola jurtina* (Lepidoptera: Satyridae). *Journal of Animal Ecology*, **42**, 773–78.

Schilthuizen, M., and Stouthamer, R. 1997. Horizontal transmission of parthenogenesis-inducing microbes in *Trichogramma* wasps. *Proceedings of the Royal Society of London, Series B*, **264**, 361–66.

Schilthuizen, M., Honda, J., and Stouthamer, R. 1998. Parthenogenesis-inducing *Wolbachia* in *Trichogramma kaykai* originates from a single infection. *Annals of the Entomological Society of America*, **91**, 410–14.

Schneider, D. 1987. The strange fate of pyrrolizidine alkaloids. In *Perspectives in Chemoreception and Behavior* (R. Chapman, E. Bernays, and J. Stoffolano, eds.), pp. 123–143. Heidelberg: Springer Verlag.

Schneider, J. M., and Elgar, M. A. 1998. Spiders hedge genetic bets. *Trends in Ecology and Evolution*, **13**, 218–219

Schwacha, A., and Kleckner, N. 1994. Identification of joint molecules that form frequently between homologs but rarely between sister chromatids during yeast meiosis. *Cell*, **76**, 51–63.

———. 1997. Interhomolog bias during meiotic recombination: meiotic functions promote a highly differentiated interhomolog-only pathway. *Cell*, **90**, 1123–35.

Seiler, J. 1959. *Untersuchungen über die Entstehung der Parthenogenese bei* Solenobia triquetrella *F.R. I. Die Zytologie der bisexuellen* Solenobia triquetrella, *ihr Verhalten und ihr Sexualverhältnis. Chromosoma*, **10**, 73–114.

———. 1960. *Untersuchungen über die Entstehung der Parthenogenese bei* Solenobia triquetrella *F.R. II. Analyse der diploid parthenogenetischen* Solenobia triquetrella. *Verhalten, Aufzuchtresultate und Zytologie. Chromosoma*, **11**, 29–102.

Shakespeare, W. 1994. *The Tragedy of Macbeth*. (Nicholas Brooke, ed., for The Oxford Shakespeare.) Oxford: Oxford University Press.

Shine, R. 1999. Why is sex determined by nest temperature in many reptiles? *Trends in Ecology and Evolution*, **14**, 186–89.

Simmonds, H. W. 1926. Sex ratio of *Hypolimnas bolina* in Viti Levu, Fiji. *Proceedings of the Royal Entomological Society of London*, **1**, 29–32.

Simmons, L. W., Stockley, P., Jackson, R. L., and Parker, G. A. 1996. Sperm competition or sperm selection: No evidence for female influence over paternity in yellow dung flies *Scatophagastercoraria*. *Behavioural Ecology and Sociobiology*, **38**, 199–206.

Siskins, S. P., Braig, H. R., and O'Neill, S. L. 1995. *Wolbachia* superinfections and the expression of cytoplasmic incompatibility. *Proceedings of the Royal Society of London, Series B*, **261**, 325–30.

Siskins, S. P., Curtis, C. F., and O'Neill, S. L. 1997. The potential application of inherited symbiont systems to pest control. In *Influential Passengers* (S. L. O'Neill, A. A. Hoffmann, and J. H. Werren, eds.), pp. 155–175.

Sivinski, J. M., and Petersson, E. 1997. Mate choice and species isolation in swarming insects. In *Mating Systems in Insects and Arachnids*. (J.C. Choe and B.J. Crespi, eds.), pp. 294–309. Cambridge: Cambridge University Press.

Skinner, S. W. 1985. Son-killer: a third extrachromosomal factor affecting sex ratios in the parasitoid wasp *Nasonia vitripennis*. *Genetics*, **109**, 745–54.

Smith, D.A.S., Gordon, I. J., Depew, L. A., and Owen, D. F. 1998. Genetics of the butterfly *Danaus chrysippus* L. in a broad hybrid zone, with special reference to sex ratio, polymorphism and intragenomic conflict. *Biological Journal of the Linnean Society*, **65**, 1–40.

Smith, J. E., and Dunn, A. M. 1991. Transovarial transmission. *Parasitology Today* **7**, 146–48.

Snook, R. R., Cleland, S. Y., Wolfner, M. F., and Karr, T. L. 2000. Offsetting effects of *Wolbachia* infection and heat shock on sperm production in *Drosophila simulans*: Analyses of fecundity, fertility and accessory gland proteins. *Genetics*, **155**, 167–78.

Snustad, D. P., Simmons, M. J., and Jenkins, J. B. 1997. *Principles of Genetics*. New York: John Wiley and Sons.

Steinemann, M., and Steinemann, S. 1992. Degenerating Y chromosome of *Drosophila miranda*: A trap for retrotransposons. *Proceedings of the National Academy of Sciences USA*, **89**, 7591–95.

Stouthamer, R. 1993. The use of sexual versus asexual wasps in biological control. *Entomophaga*, **38**, 3–6.

———. 1997. *Wolbachia*-induced parthenogenesis. In *Influential Passengers* (S. L. O'Neill, A. A. Hoffmann, and J. H. Werren, eds.), pp. 102–24. Oxford: Oxford University Press.

Stouthamer, R., and Luck, R. F. 1993. Influence of microbe-associated parthenogenesis on the fecundity of *Trichogramma deion* and *T. pretiosum*. *Entomologia Experimentalis et Applicata*, **67**, 150–57.

Stouthamer, R., and Werren, J. H. 1993. Microorganisms associated with parthenogenesis in wasps of the genus *Trichogramma*. *Journal of Invertebrate Pathology*, **61**, 6–9.

Stouthamer, R., Pinto, J. D., Platnere, G. R., and Luck, R. F. 1990a. Taxonomic status of thelytokous forms of *Trichogramma* (Hymenoptera: Trichogrammatidae). *Annals of the Entomological Society of America*, **83**, 475–81.

Stouthamer, R., Luck, R. F., and Hamilton, W. D. 1990b. Antibiotics cause parthenogenetic *Trichogramma* to revert to sex. *Proceedings of the National Academy of Sciences*, **87**, 2424–27.

Stouthamer, R., Breeuwer, J.A.J., Luck, R. F., and Werren, J. H. 1993. Molecular identification of microorganisms associated with parthenogenesis. *Nature*, **361**, 66–68.

Stouthamer, R., Breeuwer, J.A.J., and Hurst, G.D.D. 1999. *Wolbachia pipientis*: Microbial manipulator of arthropod reproduction. *Annual Review of Microbiology*, **53**, 71–102.

Sulston, J., and Hodgkin, J. 1988. Methods. In *The nematode* Caenorrhabditis elegans (W. B. Wood, ed.), pp. 587–606. Cold Spring Harbor, NY: Cold Spring Harbor Laboratory.

Taylor, D. R. 1990. Evolutionary consequences of cytoplasmic sex ratio distorters. *Evolutionary Ecology*, **4**, 235–48.

Tazima, Y. 1964. *The Genetics of the Silkworm*. London: Logos Press.

Thornhill, R. 1976. Sexual selection and nuptial feeding behaviour in *Bittacus apicalis* (Insecta: Mecoptera). *American Naturalist*, **110**, 529–48.

Torres, R., and Drummond, H. 1999. Variably male-biased sex ratio in a marine bird with females larger than males. *Oecologia*, **118**, 16–22.

Trivers, R. L. 1985. *Social Evolution*. Menlo Park, CA: Benjamin/Cummings.

Trivers, R. L., and Hare, H. 1976. Haplodiploidy and the evolution of social insects. *Science*, **191**, 249–63.

Trivers, R. L., and Willard, D. E. 1973. Natural selection of parental ability to vary the sex ratio of offspring. *Science*, **179**, 90–92.

Turelli, M., and Hoffmann, A. A. 1991. Rapid spread of an inherited incompatibility factor in California *Drosophila*. *Nature*, **353**, 440–42.

Vala, F., Breeuwer, J.A.J., and Sabelis, M. W. 2000. *Wolbachia* diversity in crustaceans. *First International Wolbachia Conference Program and Abstracts*, 34.

Vandekerckhove, T.T.M., Watteyne, S., Willems, A., Swings, J. G., Mertens, J., and Gillis, M. 1999. Phylogenetic analysis of the 16S rDNA of the cytoplasmic bacterium *Wolbachia* from the novel host *Folsomia candida* (Hexapoda, Collembola) and its implications for wolbachial taxonomy. *FEMS Microbiology Letters*, **180**, 279–86.

Vandekerckhove, T.T.M., Watteyne, S., Bocksoen, L., Swings, J. G., Mertens, J., and Gillis, M. 2000. *Wolbachia*-induced thelytoky in the springtail *Folsomia candida* (Hexapoda, Collembola). *First International Wolbachia Conference Program and Abstracts*, 35–37.

Wade, M. J., and Chang, N. W. 1995. Increased male fertility in *Tribolium confusum* beetles after infection with the intracellular parasite, *Wolbachia*. *Nature*, **373**, 72–74.

Ward, P. I. 1998. A possible explanation for cryptic female choice in the yellow dung fly, *Scathophaga stercoraria* (I.). *Ethology*, **104**, 97–110.

Warner, R. R., and Hoffman, S. G. 1980. Local population size as a determinant of mating system and sexual composition in two tropical marine fishes (*Thalassoma* spp.). *Evolution*, **34**, 508–18.

Weeks, A. R., and Breeuwer, J.A.J. 2000. What role have *Wolbachia* played in the diversification of a genus of thelytokous mite? *First International Wolbachia Conference Program and Abstracts*, 21.

Weismann, A. 1889. The significance of sexual reproduction in the theory of natural selection. In *Essays upon Heredity and Kindred Biological Problems* (E. B. Poulton, S. Schönland, and A. E. Shipley, eds.), pp. 251–332. Oxford: Clarendon Press.

Werren, J. H. 1980. Sex ratio adaptations to local mate competition in a parasitic wasp. *Science*, **208**, 1157–59.

Werren, J. H., and Bartos, J. D. 2001. Recombination in *Wolbachia*. *Current Biology*, **11**, 431–35.

Werren, J. H., and Beukeboom, L. W. 1998. Sex determination, sex ratios, and genetic conflict. *Annual Review of Ecology and Systematics*, **29**, 233–61.

Werren, J. H., and O'Neill, S. L. 1997. The evolution of heritable symbionts. In *Influen-*

tial Passengers (S. L. O'Neill, A. A. Hoffmann, and J. H. Werren, eds.), pp. 1–41. Oxford: Oxford University Press.

Werren, J. H., and Windsor, D. M. 2000. *Wolbachia* infection frequencies in insects: Evidence of a global equilibrium? *Proceedings of the Royal Society of London, Series B*, **267**, 1277–85.

Werren, J. H., Skinner, S. W., and Charnov, E. L. 1981. Paternal inheritance of a daughterless SR factor. *Nature*, **293**, 467–68.

Werren, J. H., Nur, U., and Wu, C-I. 1988. Selfish genetic elements. *Trends in Ecology and Evolution*, **3**, 297–302.

Werren, J. H., Hurst, G.D.D., Zhang, W., Breeuwer, J.A.J., Stouthamer, R., and Majerus, M.E.N. 1994. Rickettsial relative associated with male killing in the ladybird beetle (*Adalia bipunctata*). *Journal of Bacteriology*, **176**, 388–94.

Werren, J. H., Windsor, D., and Guo, L.R. 1995a. Distribution of *Wolbachia* among neotropical arthropods. *Proceedings of the Royal Society of London, Series B*, **261**, 55–63.

Werren, J. H., Zhang, W., and Guo, L. R. 1995b. Evolution and phylogeny of *Wolbachia*-reproductive parasites of arthropods. *Proceedings of the Royal Society of London, Series B*, **261**, 55–63.

West, S. A., and Herre, E. A. 1998. Partial local mate competition and sex ratio: A study on non-pollinating fig wasps. *Journal of Evolutionary Biology*, **11**, 531–48.

West, S. A., Cook, J. M., Werren, J. H., and Godfray, H.C.J. 1998. *Wolbachia* in two insect host-parasitoid communities. *Molecular Ecology*, **7**, 1457–65.

West, S. A., Lively, C. M., and Read, A. F.1999 A pluralist approach to sex and recombination. *Journal of Evolutionary Biology*, **12**, 1003–12.

Wilson, F., and Woodcock, L. T. 1960a. Environmental determination of sex in a parthenogenetic parasite. *Nature*, **186**, 99–100.

———. 1960b. Temperature determination of sex in a parthenogenetic parasite, *Ooencyrtus submettalicu*. *Australian Journal of Zoology*, **8**, 153–69.

Wyndham, J. 1965. *Consider Her Ways and Other Stories*. London: Penguin.

Wratten, S. D. 1976. Searching by *Adalia bipunctata* (L.) (Coleoptera:Coccinellidae) and escape behaviour of its aphid and ciccadellid prey on lime (*Tilia* × *vulgaris* Hayne). *Ecological Entomology*, **1**, 139–42.

Yasui, Y. 1998. The "genetic benefits" of female multiple mating reconsidered. *Trends in Ecology and Evolution*, **13**, 246–50.

Yen, J. H., and Barr, A. R. 1973. The etiological agent of cytoplasmic incompatibility in *Culex pipiens*. *Journal of Invertebrate Pathology*, **22**, 242–50.

———. 1974. Incompatibility in *Culex pipiens*. In *The Use of Genetics in Insect Control* (R. Pal and J. Whitten, eds.), pp. 97–118. Amsterdam: Elsevier.

Yusuf, M., Turner, B., Whitfield, P., Miles, R., and Pacey, J. 1999. Electron microscopical evidence of a vertically transmitted *Wolbachia*-like parasite in the parthenogenetic, stored-product pest *Liposcelis bostrychophila* Badonnel (Psocoptera). *Journal of Stored Product Research*, **36**, 169–75.

Zahavi, A. 1975. Mate selection—Selection for a handicap. *Journal of Theoretical Biology*, **67**, 603–5.

INDEX